Recent applications of generalized inverses

S L Campbell (Editor)
North Carolina State University

Recent applications of generalized inverses

π

Pitman Advanced Publishing Program
BOSTON · LONDON · MELBOURNE

PITMAN BOOKS LIMITED
128 Long Acre, London WC2E 9AN

PITMAN PUBLISHING INC
1020 Plain Street, Marshfield, Massachusetts

Associated Companies
Pitman Publishing Pty Ltd, Melbourne
Pitman Publishing New Zealand Ltd, Wellington
Copp Clark Pitman, Toronto

AMS Subject Classifications: (main) 15–02, 15A09, 47A99
(subsidiary) 47–02, 65F10, 15A60

Library of Congress Cataloging in Publication Data

Main entry under title:

Recent applications of generalized inverses.

(Research notes in mathematics; 66)
1. Matrix inversion. I. Campbell, S. L.
(Stephen La Vern) II. Title.
QA188.R4 512.9′434 82-7510
ISBN 0-273-08550-6 AACR2

British Library Cataloguing in Publication Data

Recent applications of generalized inverses.
—(Research notes in mathematics; 66)
1. Matrix inversion
I. Campbell, S. L. II. Series
512.9′434 QA188

ISBN 0-273-08550-6

Reproduced and printed by photolithography
in Great Britain by Biddles Ltd, Guildford

ISBN 0 273 08550 6

Contents

1 Introduction

S L CAMPBELL

Recent applications of generalized inverses

1. INTRODUCTION

This volume had its genesis at a special section during the 1976 AMS Regional Conference held at Columbia, South Carolina. This section was set up by Muir Z. Nashed of the University of Delaware, Newark, Delaware, and chaired by Carl D. Meyer, Jr., of North Carolina State University, Raleigh, North Carolina.

Originally, there was to be a published conference proceedings. Between the time of the Columbia meeting and its appearance in 1982, the volume's rationale, purpose, content, and editorship have changed. The remainder of this section will discuss the scope and intent of this volume in its current form. The remainder of this paper will discuss how the papers in this volume fit in with the current state of the theory of generalized inverses. Some reference will be made to recent work not covered by this volume. The referencing is not intended to be complete. The reader interested in further study of any of these topics is referred to the bibliographies of the cited papers. Cited papers which appear in this volume are denoted by double brackets[[ ]].

There exist several volumes on generalized inverses prior to 1976, (see for example) [1], [6], [10], [12], [13], [54], [60]. In 1976 there appeared an excellent and extensive survey volume [53] with an almost exhaustive bibliography.

The mid-1970's have seen somewhat of a change in the direction and type of research done on and with generalized inverses. Prior to this period, research was often concerned with equation solving ((1) - inverses) and least squares inverses. This reflects, in part, the original impetous for the study of generalized inverses in statistics. The relationship between generalized inverses and statistics is still of some interest [56]. However, increasing amounts of research have been done on such topics as; infinite dimensional theory, numerical considerations, matrices of special type (boolean, integral), matrices over algebraic structures other than the real or complex numbers, systems theory, and non-equation solving inverses. Of course, all of these topics have their roots in the 1960's and some, such as the first two, are well represented in [12], [53]. The point is that the current flavor of research on generalized inverses has changed.

It is the intent of this volume to capture this "flavor" by publishing original, "state of the art" papers that will not appear elsewhere, the majority of which relate to "recent applications". Some of the original papers presented at the 1976 conference were judged to be still current and have been included in their original form. Any that may have become dated since that conference have been revised as of the Fall of 1981. In addition, new papers were solicited during 1981 to fill gaps in this volume's coverage.

Traditional applications, such as least squares analysis in finite dimensional spaces are well covered in earlier volumes, and will not be repeated here though interest continues [38], [48]. This volume is designed to complement and update earlier works and not to supplement or replicate them. The coverage is, of course, not exhaustive and is reflective of both interests of the editor and the requirement that the work not appear elsewhere.

Finally, it should be pointed out that generalized inverses, like the ordinary inverse, are frequently not essential to the development. However, they often provide a clarifying conceptual tool that adds insight and simplifies the development. There are two exceptions the author is aware of where generalized inverses provide simple expressions for projections for which alternative simple algebraic formula do not seem to exist. One is the characterization of the orthogonal projection onto the intersection of two subspaces due to Anderson and discussed in [5]. The second is the characterization of the subspace of consistent initial conditions for a singular linear system of differential equations as $\hat{A}^D\hat{A}$ which is due to Campbell, Meyer, and Rose [14]. Related to this last application is the work of Meyer on Markov Chains [14], [48], [[50]], [51].

In many problems involving singular behavior of some kind, the ideas of a generalized inverse are implicit. This volume is concerned only in those developments where the generalized inverse appears explicitly and plays a major role.

2. DIFFERENTIAL AND DIFFERENCE EQUATIONS

One area of current research on the applications of generalized inverses deals with singular systems of (nonlinear) differential equations of the form

$$\underline{A}\dot{\underline{x}} + \underline{B}(\underline{x}) = \underline{f}. \tag{1}$$

Such equations arise in singular perturbations, cheap control problems, and descriptor systems. The basic material on the application of the Drazin inverse to (1) may be found in [14], [16]. In this work the Drazin inverse makes possible a simple characterization of certain subspaces which would ordinarily be defined by a iterative procedure.

Additional results and a fairly complete bibliography may be found in [17] which will appear in 1982. Accordingly this topic will not be explicitly covered in this volume. However, the papers of Campbell [[18]] and Bouldin [[11]] do discuss some of the problems in extending the results of [14], [16] to infinite dimensional spaces. Wilkinson's paper [[69]] explains how one might actually implement these ideas in solving linear systems in the form (1). He also discusses how one can compute the Drazin inverse.

The equation solving inverses ((1) - inverses and Moore-Penrose inverses) are sometimes used in the context of differential equations, usually in the systems and control literature. See [42],[45] for example, and [16], [17], [19], [25], [46], [47].

Differential equations can also be viewed as operator equations in infinite dimensional spaces. This will be commented on shortly.

3. ITERATIVE PROCESSES AND NONNEGATIVITY

In [48], (see also [14]) Meyer showed how the group inverse (a special case of the Drazin inverse) could be used to simplify the study of Markov chains. This approach was applied to error estimates in [49], [51]. New results may be found in Meyer's paper [[50]].

With the increased study and concern with large scale linear systems, iterative procedures based on splittings have once again become important. The group inverse was used by Meyer and Plemmons in [52] to study singular splittings. The paper by Hartwig and Hall [[35]] on Cesaro-Neumann iterations is an extension of this earlier work.

Around this same time the volume by Berman and Plemmons on nonnegative matrices [8] came out. This work, and in particular [8] and [52], has spawned a great deal of very recent work dealing with generalized inverses.

Some has dealt directly with iterative procedures [15], [20], [55], [61]. Many papers have been written about generalized inverses of particular types of matrices; circulant [4], [70] incidence [9], [38], integral [26], boolean [4], banded [7], non negative [34], [37], [39], [58], [66], [67], and polynomial or rational [63], [65]. A recent survey on non negative matrices is given in [67]. Closely related is the idea of extending the theory to matrices over finite fields and rings [23], [[35]], [36], [63]. The Drazin inverse for a matrix over a finite field [36] has been applied to certain Cryptographic systems [43]. The Drazin inverse is only defined for square matrices. Since rectangular systems of differential equations arise, a Drazin inverse for a rectangular matrix has been defined and studied [24].

4. NUMERICAL PROBLEMS

Generalized inverses arise in several ways in numerical analysis.

The first, and most obvious question, is how to (and what it means to) compute the different generalized inverses. A closely related problem is how to compute the solutions or expressions involving generalized inverses. The situation is similar to finding A^{-1} and solving $A\underline{x} = \underline{b}$ for nonsingular A. Both problems involve similar operations but one would not normally solve $A\underline{x} = \underline{b}$ by computing A^{-1} and then $A^{-1}\underline{b}$. In this volume Wilkinson [[69]] discusses both types of problems in relation to the Drazin inverse and solutions of linear systems of differential equations. The comparable results for the Moore-Penrose and (1) - inverses in the finite dimensional case were known prior to 1976. Some work is still underway [26], [40]. See [64] for a good summary. Generalized inverses can also be used to study the behavior of ill conditioned systems.

Generalized inverses can be part of the procedure for numerically solving a problem. One example was the singular splittings discussed in Section 3, (see [55] for example). Another is Tanabe's paper [[68]] in this volume which shows how in some minimization problems even if the appropriate Jacobian is singular, a generalized inverse can still often be used in a Newton type procedure. See also [27]. As to be expected, in the infinite dimensional case the situation is sometimes more complicated but similar developments exist.

5. ELECTRICAL NETWORKS

Of course the work mentioned in Section 2 may be applied to electrical networks. As noted in [17], certain important stabilizable and reachable subspaces for singular systems of the form (1) can be described using the Drazin inverse.

One of the more original applications involving generalized inverses has been the work of Trapp, Anderson, Morley and Duffin in network connections and shorted operators. Some of their early results may be found in [14]. The papers [[2]], [[29]] extend and unify this earlier work.

6. INFINITE DIMENSIONAL PROBLEMS

The study of generalized inverses in infinite dimensional spaces, as to be expected, has several different aspects.

First one needs to distinguish between infinite matrices and linear operators on some type of topological vector space. Infinite matrices, for which multiplication can be nonassociative, are considered in [17, Chapter VII] and [[18]]. Most of the literature, however, has assumed some type of topology, and in fact, has tended to be in Hilbert (complete, inner-product), or Banach (complete normed linear) spaces.

There exists a large body of work on defining various types of generalized inverses, determining when they exist, developing their basic properties, and applying them. See, for example [3], [12], [19], [21], [53], [54]. The last five papers in this volume continue these studies.

The spectral decomposition of a Hermitian matrix plays an important role in the theory of the Moore-Penrose inverse and the singular value decomposition. Decomposable operators on Banach spaces represent a generalization of the spectral theory for Hermitian matrices. In this volume, Erdelyi [[31]] studies the generalized inversion of decomposable operators. Campbell [[18]] and Bouldin [[11]] discuss some of the problems of extending the theory of [14], [16], [17] to infinite systems of differential equations and denumberable Markov chains.

Another important question is the calculation of specific generalized inverses both to serve as examples and applications of the theory. Differential operators are still studied [30], [44] and have been recently applied to bifurcation theory [44]. In this volume, Groetsch and Jacobs [[32]] develop an iterative method to compute generalized inverses and Chang [[22]]

applies the theory of generalized inverses to interpolation theory.

Much of the work in the papers of Anderson and Trapp [[2]] and Duffin and Morley [[29]] mentioned earlier can be placed in an infinite dimensional setting.

REFERENCES

1. A. Albert, Regression and the Moore-Penrose Pseudoinverse, Academic Press, New York, 1972., 180 pp.

2. W. N. Anderson and G. E. Trapp, Analytic operator functions and electrical networks, This Volume.

3. P. M. Anselone and M. Z. Nashed, Perturbation of outer inverses, Approximation Theory III, (Edited by E. W. Cheney), Academic Press, New York, 1980, 163-169.

4. C. L. Bell, Generalized inverses of circulant and generalized circulant matrices, Linear Alg. and Its Appl., 39 (1981), 133-142.

5. A. Ben-Israel, Generalized inverses of matrices and their applications, Lecture Notes in Economics and Mathematical Systems, Vol. 174, Springer-Verlag, 1980, 154-186.

6. A. Ben-Israel and T. N. E. Greville, Generalized Inverses: Theory and Applications, Wiley-Interscience, New York, 1974, xi + 395 pp.

7. L. Berg, Asymptotic evaluation of the Moore-Penrose inverses of rectangular band matrices, ZAMM 59 (1979), 317-319.

8. A. Berman and R. J. Plemmons, Nonnegative Matrices in the Mathematical Sciences, Academic Press, New York, 1979.

9. J. H. Bevis, F. J. Hall and I. J. Katz, Integer generalized inverses of incidence matrices, Linear Alg. and Its Appl., 39(1981), 247-258.

10. A. Bjerhammar, A Theory of Errors and Generalized Inverse Matrices, Elsevier, Amsterdam, 1973, 440 pp.

11. R. H. Bouldin, Generalized inverses and factorizations, This Volume.

12. T. L. Boullion and P. L. Odell (Editors), Theory and Applications of Generalized Inverses of Matrices, Texas Tech. Press, Lubbock, Texas, 1969.

13. T. L. Boullion and P. L. Odell, Generalized Inverse Matrices, Wiley-Intersciences, New York, 1971, 108 pp.

14. S. L. Campbell and C. D. Meyer, Jr., Generalized Inverses of Linear Transformations, Pitman, London, 1979.

15. S. L. Campbell, Limit behavior of solutions of singular difference equations, Linear Alg. and Its Appl., 23(1979), 167-178.

16. S. L. Campbell, Singular Systems of Differential Equations, Pitman, London, 1980.

17. S. L. Campbell, Singular Systems of Differential Equations II, Pitman, London, 1982 (In Press).

18. S. L. Campbell, The Drazin inverse of an operator, This Volume.

19. S. L. Campbell and G. D. Faulkner, Operators on Banach spaces with complemented ranges, Acta Math. Acad. Sci. Hung. 35(1980), 123-128.

20. S. L. Campbell and G. D. Poole, Convergent regular splittings for nonnegative matrices, Linear and Multilinear Alg. 10(1981), 63-73.

21. S. R. Caradus, Generalized Inverses and Operator Theory, Queens University Press, 1978.

22. E. Chang, The generalized inverse and interpolation theory, This Volume.

23. R. E. Cline, Note on an extension of the Moore-Penrose inverse, Linear Alg. and Its Appl., 40(1981), 19-24.

24. R. E. Cline and T. N. E. Greville, A Drazin inverse for rectangular matrices, Linear Alg. and Its Appl., 29(1980), 53-62.

25. B. Das and T. K. Ghoshal, Reduced-order observer construction by generalized matrix inverse, International J. Control 33(1981), 371-378.

26. P. Deuflhard and W. Sautter, On rank-deficient pseudoinverses, Linear Alg. and Its Appl., 29(1980), 91-111.

27. P. Deuflhard and V. Apostolescu, A study of the Gauss-Newton algorithm for the solution of nonlinear least squares problems, preprint, 1980.

28. M. P. Drazin, Differentiation of generalized inverses, This Volume.

29. R. J. Duffin and T. D. Morley, Inequalities induced by network connections, This Volume.

30. H. W. Engl and M. Z. Nashed, Generalized inverses of random linear operators in Banach spaces, J. Math. Anal. Appl., 83(1981), 582-610.

31. I. Erdelyi, Spectral decompositions for generalized inversions, This Volume.

32. C. W. Groetsch and B. J. Jacobs, Iterative methods for generalized inverses based on functional interpolation, This Volume.

33. F. J. Hall and I. J. Katz, More on integral generalized inverses of integral matrices, Linear and Multilinear Alg., 9(1980), 201-209.

34. F. J. Hall and I. J. Katz, Nonnegative integral generalized inverses, Linear Alg. and Its Appl., 39(1981), 23-40.

35. R. E. Hartwig and F. J. Hall, Applications of the Drazin Inverse to Cesaro-Neumann Iterations, This Volume.

36. R. E. Hartwig, Drazin inverses and canonical forms in $M_n(Z/h)$, Linear Alg. and Its Appl., 37(1981), 205-233.

37. E. Haynsworth and J. R. Wall, Group inverses of certain nonnegative matrices, Linear Alg. and Its Appl., 25(1979), 271-288.

38. Y. Ijiri, Cost-flow networks and generalized inverses, Lecture Notes in Economics and Mathematical Systems, Vol. 174, Springer-Verlag, 1980, 187-196.

39. S. K. Jain and L. E. Snyder, Nonnegative λ-monotone matrices, SIAM J. Alg. Disc. Methods, 2(1981), 66-76.

40. K. O. Kymn, J. R. Norsworthy, and T. Okamoto, The computation of the generalized inverse, Kyungpook Math. J. 20(1980), 199-206.

41. W. F. Langford, The generalized inverse of an unbounded linear operator with unbounded constraints, SIAM J. Math. Anal., 9(1978), 1083-1095.

42. F. Lewis, A generalized inverse solution to the discrete-time singular Riccati equation, IEEE Trans. Automatic Control, AC-26(1981), 395-398.

43. J. Levine and R. E. Hartwig, Applications of the Drazin inverse to the Hill cryptographic system, I. Cryptologia, 4(1980), 71-83.

44. W. S. Loud, A bifurcation application of the generalized inverse of a linear differential operator, SIAM J. Math. Anal., 3(1980), 545-558.

45. V. Lovass-Nagy, R. J. Miller, and R. Mukundan, On the application of matrix generalized inverses to the design of observers for time-varying and time-invariant linear systems, IEEE Trans. Automatic Control, AC-25(1980), 1213-1218.

46. V. Lovass-Nagy and R. Mukundan, Determination of time-exponential inputs generating time-exponential state responses for linear time-invariant control systems, International J. Control, 31(1980), 153-158.

47. V. Lovass-Nagy and D. L. Powers, Determination of zeros and zero directions of linear time-invariant systems by matrix generalized inverses, International J. Control, 31(1980),1161-1170.

48. C. D. Meyer, Jr., The role of the group generalized inverse in the theory of finite Markov chains, SIAM Review, 17(1975), 443-464.

49. C. D. Meyer, Jr., The condition of a finite Markov chain and perturbation bounds for the limiting probabilities, SIAM J. Alg. Disc. Methods, 1(1980), 273-283.

50. C. D. Meyer, Jr., Analysis of finite Markov chains by group inversion techniques, This Volume.

51. C. D. Meyer, Jr., Differentation of the limiting distribution for ergodic Markov chains, preprint, 1982.

52. C. D. Meyer, Jr., and R. J. Plemmons, Convergent powers of a matrix with applications to iterative methods for singular linear systems, SIAM J. Numer. Anal., 14(1977), 699-705.

53. M. Z. Nashed (Editor), Generalized Inverses and Applications, Academic Press, New York, 1976, xiv + 1054 pp.

54. M. Z. Nashed, Generalized inverse mapping theorems and related applications of generalized inverses in nonlinear analysis, in Nonlinear Equations in Abstract Spaces, Proceedings International Symposium, Arlington, 1977 (1978), 217-252.

55. M. E. Neumann, A combined direct-iterative approach for solving large scale singular and rectangular consistent systems of linear equations, Linear Alg. and Its Appl., 34(1980), 85-101.

56. D. V. Ouellette, Schur complements and statistics, Linear Alg. and Its Appl., 36(1981), 187-195.

57. R. M. Pringle and A. A. Rayner, Generalized Inverse Matrices with Applications to Statistics, Griffin, London, 1971.

58. W. C. Pye, Nonnegative Drazin inverses, Linear Alg. and Its Appl., 30(1980), 149-153.

59. L. D. Pyle, The weighted generalized inverse in nonlinear programming-active set selection using a variable-metric generalization of the simplex algorithm, Lecture Notes in Economics and Mathematical Systems, Col. 174, Springer-Verlag, 1980, 197-230.

60. C. R. Rao and S. K. Mitra, Generalized Inverse of Matrices and its Applications, Wiley, New York, 1971, xiv + 240 pp.

61. U. G. Rothblum, Convergence properties of powers of matrices with applications to iterative methods for solving linear systems, Lecture Notes in Economics and Mathematical Systems, Vol. 174, Springer-Verlag, 1980, 231-247.

62. A. Smeds, Line digraphs and the Moore-Penrose inverse, Linear Alg. and Its Appl., 36(1981), 165-172.

63. E. D. Sontag, On generalized inverses of polynomial and other matrices, IEEE Trans. Automatic Control, AC-25(1980), 514-517.

64. G. W. Stewart, On the perturbations of pseudo-inverses, projections and linear least squares problems, SIAM Review, 19(1977), 634-662.

65. M. G. Striutzis, Diagonalization, rank calculation and generalized inverses of rational matrices, Applied Math. and Computation 3(1977), 68-94.

66. B.-S. Tam, Generalized inverses of cone preserving maps, Linear Alg. and Its Appl., 40(1981), 189-202.

67. B.-S. Tam, A geometric treatment of generalized inverses and semigroups of nonnegative matrices, Linear Alg. and Its Appl., 41(1981), 225-272.

68. K. Tanabe, Differential geometric approach to extended GRG methods with enforced feasibility in nonlinear programming: Global analysis, This Volume.

69. J. H. Wilkinson, Note on the practical significance of the Drazin inverse, This Volume.

70. E. T. Wong, Polygons, circulant matrices, and Moore-Penrose inverses, Amer. Math. Monthly 88 (1981), 509-515.

S. L. Campbell
Department of Mathematics
North Carolina State University
Raleigh, North Carolina 27650

The rapid and efficient typing by Donna Ward and Sharon Jones is gratefully acknowledged.

W N ANDERSON AND G E TRAPP

Analytic operator functions and electrical networks

1. INTRODUCTION

One important application of operator valued analytic functions is to the study of electrical networks; in this tutorial paper we will discuss some aspects of this application. Among the standard references for the results presented here are the textbooks of Hazony [14] and Newcomb [16]. Our particular treatment is based on the papers of Anderson, Duffin and Trapp [1], Bott and Duffin [7] and Duffin [12]. Our notation does not follow the electrical network theory standard.

In Section 2 we discuss the concept of a positive real operator, the type of operator function which arises naturally in studying electrical networks. In Section 3 we discuss the algebraic setting for the study of electrical networks. In Section 4 we show how n-port electrical networks can be analyzed using the techniques previously developed. In the final section we discuss some other aspects of the theory.

In Sections 2, 3, and 4 we will be dealing with linear operators defined on finite dimensional complex Hilbert spaces; inner products will be denoted by $\langle \cdot,\cdot \rangle$. We will let E and P denote respectively m and n dimensional spaces. We assume that E has a fixed orthonormal basis $\{e_1,\ldots,e_m\}$, whose members are called edges, and that P has a fixed orthonormal basis $\{p_1,\ldots,p_n\}$, whose members are called ports. All matrices will be written with respect to these bases. The letter λ will denote a complex number; the letter ω will always be used for a real number. For a linear operator A the adjoint operator A^* is defined by $\langle Ax,y \rangle = \langle x,A^*y \rangle$ for all vectors x and y. If $A = A^*$, then we say that A is Hermitian. If A is Hermitian, then we say that A is positive semidefinite if $\langle Ax,x \rangle \geq 0$ for all vectors x. For Hermitian operators A and B, we write $A \geq B$ if $A - B$ is positive semidefinite.

2. POSITIVE REAL FUNCTIONS

Among the most important operator valued analytic functions are the positive real operators. As we shall see in the sequel, these functions are of primary importance in electrical network theory. They also arise in many other physical and non-physical applications [5], [15], [21]. In this section we derive some of the simpler properties of positive real functions; more exhaustive treatments are contained in [14], [16], [17], [20], [22].

The linear operator A is called almost right definite if

(i) $\mathrm{Re}\ \langle Ax,x\rangle \geq 0$ for all vectors x, i.e. A is non-dissipative;

(ii) $\mathrm{Re}\ \langle Ax,x\rangle = 0$ only if $Ax = 0$.

Lemma 2.1 Let A be a linear operator with $A = A_H + A_S$, where A_H is Hermitian and A_S is skew Hermitian. Then A is almost right definite if and only if

(iii) A_H is positive semidefinite

(iv) $\mathrm{ran}(A_S) \subset \mathrm{ran}(A_H)$, where $\mathrm{ran}(A)$ denotes the range of A.

Proof It is easy to see that $\mathrm{Re}\ \langle Ax,x\rangle = \langle A_H x,x\rangle$ so that (i) and (iii) are equivalent. Now suppose that (iii) and (iv) hold, and that $\mathrm{Re}\ \langle Ax,x\rangle = 0$. Then $\langle A_H x,x\rangle = 0$, and since A_H is positive semi-definite, it follows that $A_H x = 0$. Therefore $A_S x = 0$, and thus $Ax = 0$. Conversely, suppose that (i) and (ii) hold. Then if $\mathrm{ran}(A_S) \not\subset \mathrm{ran}(A_H)$, there is a vector x such that $A_H x = 0$ but $A_S x \neq 0$. Then $\mathrm{Re}\ \langle Ax,x\rangle = \langle A_H x,x\rangle = 0$, but $Ax \neq 0$, contradicting (ii). □

It is an easy consequence of the lemma that if A is almost right definite, then $\mathrm{ran}(A) = \mathrm{ran}(A_H)$ and $\mathrm{ran}(A) = \ker(A)^{\perp}$. It follows that $\tilde{A} = A|\mathrm{ran}(A)$ maps $\mathrm{ran}(A)$ one-to-one onto $\mathrm{ran}(A)$, and thus is invertible. The operator $A^{\dagger}$, the Moore-Penrose generalized inverse of A, is defined by

$$A^{\dagger} \mid \mathrm{ran}(A) = \tilde{A}^{-1}$$

$$A^{\dagger} \mid \ker(A) = 0.$$

Lemma 2.2 If A is an almost right definite operator, then $A^{\dagger}$ is almost right definite.

Proof Let P be the Hermitian projection onto $\operatorname{ran}(A)$. For any vector x, there is a vector $y \in \operatorname{ran}(A)$ such that $Ay = Px$. Then $\langle A^{\dagger}x,x\rangle = \langle A^{\dagger}Px,Px\rangle = \langle y,Ay\rangle$. Therefore $\operatorname{Re}\langle A^{\dagger}x,x\rangle = \operatorname{Re}\langle y,Ay\rangle =$ $\operatorname{Re}\langle Ay,y\rangle \geq 0$. Moreover, if $\operatorname{Re}\langle A^{\dagger}x,x\rangle = 0$, then $0 = Ay = Px$, so that $x \in \ker(P) = \ker(A)$. Therefore $A^{\dagger}$ is almost right definite. □

The generalized inverse we have constructed above is commonly known as the Moore-Penrose generalized inverse. For many of our purposes it would suffice to consider any oeprator A' such that $AA'A = A$; our $A^{\dagger}$ is such an operator, but there exist in general many others which are not almost right definite.

Let $\mathcal{D}$ denote the open right half-plane of the complex plane. A positive real operator $A(\lambda)$ is an operator valued function defined on $\mathcal{D}$ such that

(i) $A(\lambda)$ is analytic on $\mathcal{D}$.

(ii) $A(\lambda)$ is Hermitian on the real axis.

(iii) $\operatorname{Re}\langle A(\lambda)x,x\rangle \geq 0$ for all vectors x and all $\lambda \in \mathcal{D}$

Lemma 2.3 Let $A(\lambda)$ be an analytic operator function on $\mathcal{D}$ such that $A(\lambda)$ is Hermitian on the real axis. Then $A(\lambda)$ is a positive real operator if and only if $A(\lambda)$ is an almost right definite operator for all $\lambda \in \mathcal{D}$.

Proof If $A(\lambda)$ is almost right definite for all $\lambda \in \mathcal{D}$, then (iii) holds. Conversely, suppose that $\operatorname{Re}\langle A(\lambda_o)x,x\rangle = 0$ for some $\lambda_o \in \mathcal{D}$ and vector $x \neq 0$. Consider the scalar functions $f(\lambda) = \langle A(\lambda)x,x\rangle$; then $\operatorname{Re} f(\lambda_o) = 0$. Since $\operatorname{Re} f(\lambda)$ is a harmonic function, it follows that either $\operatorname{Re}(f) \equiv 0$ on $\mathcal{D}$ or f is negative at points arbitrarily close to λ_o. The latter condition violates the positive reality of $A(\lambda)$. Therefore $\langle A(\lambda)x,x\rangle = 0$ on the real axis; since $A(\lambda)$ is Hermitian there, it follows that $A(\lambda)x = 0$ on the real axis. Now let y be any vector. Then $\langle A(\lambda)x,y\rangle = 0$ on the real axis, and therefore throughout $\mathcal{D}$ be analytic continuation. Therefore $A(\lambda)x = 0$ throughout $\mathcal{D}$; in particular $A(\lambda_o)x = 0$. □

Lemma 2.4 Let $A(\lambda)$ be a positive real operator. Then $A^{\dagger}(\lambda)$ is a positive real operator.

Proof In view of Lemmas 2.2 and 2.3, we need only show that $A^{\dagger}$ is an analytic function. It follows from the proof of Lemma 2.3 that ker $A(\lambda)$ is a constant subspace for $\lambda \in \mathcal{D}$. Therefore $A(\lambda) = \tilde{A}(\lambda) \oplus 0$, where $\tilde{A}(\lambda)$ is invertible, and $A^{\dagger}(\lambda) = \tilde{A}^{-1}(\lambda) \oplus 0$. Since a true inverse is analytic, so is $A^{\dagger}$. □

It is not true that for a general analytic operator function the Moore-Penrose generalized inverse is analytic; however, as Bart, Kaashoek and Lay have shown in [3], an analytic pseudoinverse A' satisfying $AA'A = A$ can always be constructed. For most of our needs, any such pseudoinverse will suffice.

3. CONFLUENCES

In this section we present an algebraic setting which will enable us to analyze the behavior of electrical networks. No specific reference is made to the networks themselves; the same structure may be found in non-electrical situations [7].

A confluence is a subspace $\Gamma \subset E \oplus P$ such that

(i) for each vector $p \in P$ there is a vector $e \in E$ such that $(e,p) \in \Gamma$

(ii) if $(0,p) \in \Gamma$, then $p = 0$.

If Γ is a confluence, then we define the dual confluence Γ' by
$\Gamma' = \{(\acute{\varepsilon},\acute{\pi}) \in V \mid \langle \varepsilon', e\rangle = \langle \pi', p\rangle \text{ for all } (e,p) \in \Gamma\}$.

Lemma 3.1 If Γ is a confluence, then

$$\Gamma'' = \Gamma \tag{a}$$

$$\dim \Gamma + \dim \Gamma' = \dim (E \oplus P) \tag{b}$$

Γ' is a confluence. (c)

The proof is an easy exercise in linear algebra and is omitted.

It will be convenient to represent Γ by certain matrices; recall that in this regard fixed orthonormal bases have been chosen for S and T.

Lemma 3.2 There exists an $n \times m$ matrix J and an $r \times m$ matrix K such that the columns of the matrix

$$J_\Gamma = \begin{bmatrix} J^* & K^* \\ I & 0 \end{bmatrix}$$

form a basis for Γ. Moreover $r \leq n$.

Proof By condition (i) of the definition of confluence, we can choose the first n columns of J_Γ as shown; the remaining columns are chosen to be a basis for the set of vectors in Γ which are of the form $(e,0)$. The inequality $r \leq n$ follows from parts b and c of Lemma 3.1. □

In a similar manner, we let the columns of the matrix

$$J_{\Gamma'} = \begin{bmatrix} M^* & N^* \\ I & 0 \end{bmatrix}$$

be a basis for Γ'.

Lemma 3.3 The vector $(e,p) \in \Gamma$ if and only if

$\begin{bmatrix} M \\ N \end{bmatrix} e = \begin{bmatrix} p \\ 0 \end{bmatrix}$ if and only if there is a vector θ such that

$[J^* \quad K^*] \begin{bmatrix} p \\ \theta \end{bmatrix} = e$. The vector $(\varepsilon,\pi) \in \Gamma'$ if and only if

$\begin{bmatrix} J \\ K \end{bmatrix} \varepsilon = \begin{bmatrix} \pi \\ 0 \end{bmatrix}$ if and only if there is a vector μ such that

$[M^* \quad N^*] \begin{bmatrix} \pi \\ \mu \end{bmatrix} = \varepsilon$.

The proof follows by direct computations from the definitions of the matrices J_Γ and $J_{\Gamma'}$.

The next theorem is the fundamental existence theorem which will guarantee solutions for electrical networks.

Theorem 3.4 Let Γ be a confluence and W an almost right definite operator on E. Then given a vector $p \in P$ there exists a vector $e \in E$ and a unique vector $\pi \in P$ such that $(e,p) \in \Gamma$ and $(We,\pi) \in \Gamma'$.

Proof Using Lemma 3.3, we wish to solve the equations

$$\begin{bmatrix} 0 & 0 & M \\ 0 & 0 & N \\ M^* & N^* & -W \end{bmatrix} \begin{bmatrix} \pi \\ \theta \\ e \end{bmatrix} = \begin{bmatrix} p \\ 0 \\ 0 \end{bmatrix} . \qquad (1)$$

The homogeneous adjoint system to (1) is

$$\begin{bmatrix} 0 & 0 & M \\ 0 & 0 & N \\ M^* & N^* & -W^* \end{bmatrix} \begin{bmatrix} u_1 \\ u_2 \\ u_3 \end{bmatrix} = \begin{bmatrix} 0 \\ 0 \\ 0 \end{bmatrix} . \qquad (2)$$

For any solutions to (2), we have

$$[M^* \quad N^*] \begin{bmatrix} u_1 \\ u_2 \end{bmatrix} = W^* u_3 . \qquad (3)$$

Then W^*u_3 is orthogonal to all solutions to $\begin{bmatrix} M \\ N \end{bmatrix} \alpha = 0$. Since u_3 is such a solution we have $\langle u_3, W^*u_3 \rangle = 0$. Since W is almost right definite, it follows from $W^*u_3 = 0$, so that the right-hand side of (3) is 0, and thus (3) is the homogeneous adjoint system for the system

$$\begin{bmatrix} M \\ N \end{bmatrix} e = \begin{bmatrix} p \\ 0 \end{bmatrix} . \qquad (4)$$

By the definition of confluence, system (4) has a solution for all vectors p, and thus $\langle u_1, p \rangle = 0$ for all p. Thus $u_1 = 0$, so that (1) has a solution. Moreover, since $\pi = 0$ for $p = 0$, it follows that π is uniquely determined by p. □

Since the correspondence between p and π is linear, and π is uniquely determined by p, there is a linear operator $\gamma(W)$ such that $\gamma(W)p = \pi$.

Theorem 3.5 Let W be an almost right definite operator and Γ a confluence. Then $\gamma(W)$ is an almost right definite operator.

Proof Given a vector p, there exists e_o and π such that $(e_o,p) \in \Gamma$ and $(We_o,\pi) \in \Gamma'$. Therefore $\langle We_o,e_o\rangle = \langle \pi,p\rangle = \langle \gamma(W)p,p\rangle$. Thus $\mathrm{Re}\,\langle \gamma(W)p,p\rangle \geq 0$. Now if $\mathrm{Re}\,\langle \gamma(W)p,p\rangle = 0$, then $\mathrm{Re}\,\langle We_o,e_o\rangle = 0$, so that $We_o = 0$. Since Γ is a confluence, it follows that $\pi = 0$; that is $\gamma(W)p = 0$. □

In order to investigate the properties of the function γ, it will be useful to have an explicit representation for $\gamma(W)$.

Theorem 3.6 Let Γ be a confluence and W an almost right definite operator. Then

$$\gamma(W) = JWJ^* - JWK^*(KWK^*)^{\dagger}KWJ^* \quad . \tag{5}$$

Proof By Theorem 3.4, given a vector $p \in P$ there is a vector $e \in E$ and a unique vector $\pi \in P$ such that $(e,p) \in \Gamma$ and $(We,\pi) \in \Gamma'$. Using Lemma 3.3, we find that there exists a vector θ such that

$[J^* \quad K^*] \begin{bmatrix} p \\ \theta \end{bmatrix} = e$. Moreover,

$\begin{bmatrix} J \\ K \end{bmatrix} We = \begin{bmatrix} \pi \\ 0 \end{bmatrix}$, so that

$$\begin{bmatrix} \pi \\ 0 \end{bmatrix} = \begin{bmatrix} J \\ K \end{bmatrix} W[J^* \quad K^*] \begin{bmatrix} p \\ 0 \end{bmatrix} . \tag{6}$$

In particular, $0 = KWJ^*p + KWK^*\theta$. Since θ is known to exist for any p, we may solve for one possible value of $\theta = -(KWK^*)^{\dagger}KWJ^*p$. Substituting this value for θ in (6), we have $\pi = (JWJ^* - JWK^*(KWK^*)^{\dagger}KWJ^*)p$. However, π is uniquely determined by p, so the theorem is proved. □

Theorem 3.7 Let $W(\lambda)$ be a positive real operator, and let Γ be a confluence. Then $\gamma(W(\lambda))$ is a positive real operator.

Proof Since W is positive real, so is KWK^*. Lemma 2.4 then implies that $(KWK^*)^{\dagger}$ is analytic, and thus $\gamma(W)$ is analytic. Thus Theorem 3.5 and Lemma 2.4 imply that $\gamma(W)$ is positive real. □

4. N-PORT ELECTRICAL NETWORKS

In this section we introduce the concept of an *n-port electrical network*, and show how the behavior of such a network is defined by a positive real operator.

The three basic electrical network elements are the *resistor*, the *inductor* and the *capacitor*. For each of these we will consider the *voltage* V developed across the network element by a *current* I of the form $I(t) = ae^{\lambda t}$, where t denotes time and a and λ are complex numbers. The voltage and current are then related by Ohm's law: $V(t) = Z(\lambda)I(t)$. The function $Z(\lambda)$ is known as the *impedance* of the network element. If the current is more complicated than a single exponential, then the voltage can be determined using the function $Z(\lambda)$ and Fourier analysis.

The resistor is indicated by the symbol —/\/\/\/— ; a resistor of R *Ohms* has an impedance of R *Ohms*. The inductor is indicated by the symbol —ʘʘʘ— ; and inductor of L *Henrys* has an impedance of $L\lambda$ Ohms. The capacitor is indicated by the symbol —| |— ; a capacitor of C Farads has an impedance of $1/C\lambda$ Ohms. These impedance formulas are derived in most elementary physics textbooks.

Before turning to the study of networks in general, let us first analyze a few simple examples. In the *series connection* of two impedances Z_1 and Z_2, the same current will flow through each network element, so that the voltages --

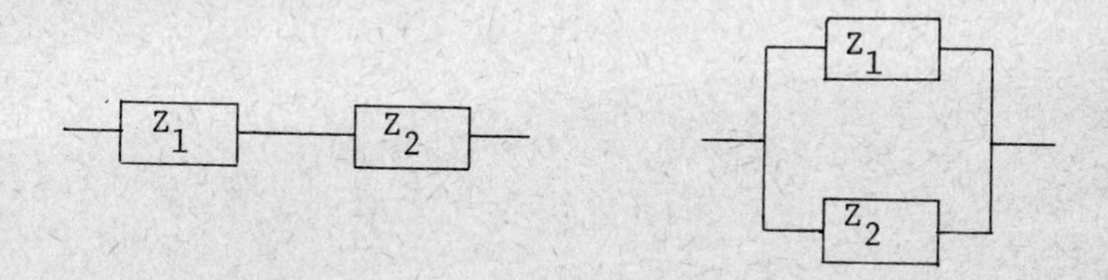

Series Connections Parallel Connections

and hence the impedances--will add. The impedance of the series connection is thus given by the formula $Z = Z_1 + Z_2$. One important special example is where $Z_1 = L\lambda$ and $Z_2 = 1/C\lambda$; then $Z = L\lambda + 1/C\lambda$. If $\lambda = i\omega$ is pure imaginary, then $Z = i(L\lambda - 1/C\lambda)$, and $Z = 0$ for $\omega = \pm 1/\sqrt{LC}$. This <u>series resonant circuit</u> is used in radios where it is desired to have a circuit which presents a high impedance to unwanted frequencies, but a low impedance to some wanted frequency.

Another important connection is the <u>parallel connection</u>. In this connection the voltage across the two network elements must be the same, and the currents add; then $Z = Z_1 Z_2/(Z_1 + Z_2)$. If $Z_1 = L\lambda$ and $Z_2 = 1/C\lambda$; then in the <u>parallel resonant</u> circuit the impedance will become infinite for $\lambda = \pm i/\sqrt{LC}$.

It may be verified that the impedance functions of series and parallel connections are positive real. We will now proceed to analyze more complicated electrical networks.

An <u>n-port electrical network</u> consists of a finite number, m, of <u>edges</u> and a finite number of <u>vertices</u>; each edge connects two distinct vertices, called the <u>head</u> and <u>tail</u> of the edge. There are also n distinct ordered pairs of vertices; these pairs are known as <u>ports</u>. Associated with a flow of current in the network, there is a vector $(e,p) \in E \oplus P$. A current of I units in the i'th edge of the network is associated with the vector $Ie_i \in E$; I will be positive or negative according as the current flows with, or against, the orientation of the edge. A current of I units into the positive vertex of the i'th port, and out of the negative vertex of the i'th port, is associated with the vector $Ip_i \in P$. For the pair (e,p), the vector e is called the <u>edge current</u> and the vector p the <u>port current vector</u>. In a similar manner we construct a vector $(\varepsilon,\pi) \in E \oplus P$ where ε is the edge voltage vector and π is the <u>port voltage vector</u>.

For each edge of the network there is an <u>edge impedance</u>. This impedance will be due to one of the three basic network elements, so that the impedance will be of the form R, $L\lambda$ or $1/C\lambda$. Let W denote the $m\times m$ diagonal matrix of edge impedances; then W is clearly positive real. W is called the edge impedance operator.

In order to analyze the behavior of the network, we need some physical assumptions about the nature of possible currents and voltages. The fundamental principles are the two laws of Kirchhoff. First, the net current that flows into any vertex must be zero. Second, the edge voltage must be due to potentials at the vertices; equivalently, the sum of the voltages around any closed circuit must be zero. In analyzing n-port networks, we wish to assume that all vectors in P are *a priori* possible for the port currents and voltages, combining this assumption with Kirchhoff's laws we obtain our

Fundamental Physical Principle: The set of possible current vectors for an n-port network forms a confluence Γ; the set of possible voltage vectors forms the dual confluence Γ'.

Given a port current vector p, we wish to determine an edge current e and the port voltage π such that $(e,p) \in \Gamma$ and $(We,\pi) \in \Gamma'$. By Theorem 3.4 these vectors exist and π is unique. The correspondence from p to π is then given by $\pi = \gamma(W)p = Z(\lambda)p$. The operator $Z(\lambda)$ is called the *port impedance operator*. Using the explicit formula (5) and Theorem 3.7 we have the fundamental result;

Theorem 4.1 The port impedance operator of an n-port network is a rational positive real operator.

We will examine here one example of a 2-port network. The reader may verify that the matrices given define a pair of dual confluences, and that the currents and voltages satisfy Kirchhoff's laws. Standard techniques are known to find these bases [14], [16]. Cederbaum's approach is closest to the idea of confluence [9].

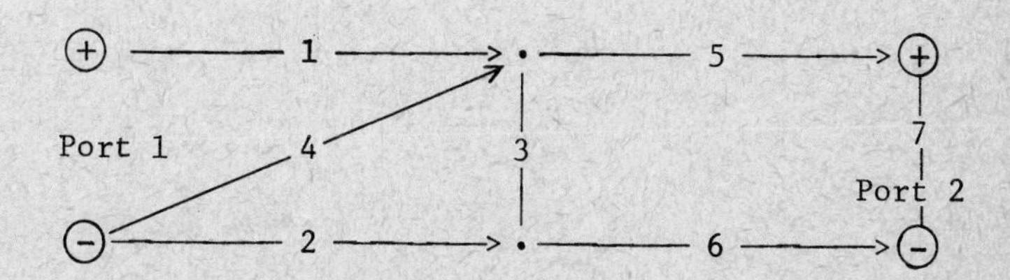

The current confluence Γ is defined by:

$$\begin{bmatrix} J^* & K^* \\ I & 0 \end{bmatrix} = \left[\begin{array}{cc|cc} 1 & 0 & 0 & 0 \\ 0 & 0 & 1 & 0 \\ 0 & 0 & -1 & 1 \\ -1 & 0 & -1 & 0 \\ 0 & 0 & 0 & -1 \\ 0 & 0 & 0 & 1 \\ 0 & -1 & 0 & 1 \\ \hline 1 & 0 & 0 & 0 \\ 0 & 1 & 0 & 0 \end{array}\right] \begin{matrix} e_1 \\ e_2 \\ e_3 \\ e_4 \\ e_5 \\ e_6 \\ e_7 \\ p_1 \\ p_2 \end{matrix}$$

The voltage confluence Γ' is defined by:

$$\begin{bmatrix} M^* & N^* \\ I & 0 \end{bmatrix} = \left[\begin{array}{cc|ccc} 1 & 0 & 1 & 0 & 1 \\ 0 & 0 & 0 & 1 & 1 \\ 0 & 0 & -1 & 1 & 0 \\ 0 & 0 & 1 & 0 & 1 \\ 0 & 0 & -1 & 0 & 0 \\ 0 & 0 & 0 & -1 & 0 \\ 0 & -1 & 0 & 0 & 0 \\ \hline 1 & 0 & 0 & 0 & 0 \\ 0 & 1 & 0 & 0 & 0 \end{array}\right] \begin{matrix} e_1 \\ e_2 \\ e_3 \\ e_4 \\ e_5 \\ e_6 \\ e_7 \\ p_1 \\ p_2 \end{matrix}$$

Now, if $W = \text{diag}(w_1,\ldots,w_7)$ we have, using (5), $Z(\lambda) =$

$$\begin{bmatrix} w_1 + w_4 & 0 \\ 0 & w_7 \end{bmatrix} - \begin{bmatrix} w_4 & 0 \\ 0 & -w_7 \end{bmatrix} \begin{bmatrix} w_2 + w_3 + w_4 & -w_5 \\ -w_3 & w_3+w_5+w_6+w_7 \end{bmatrix}^{\dagger} \begin{bmatrix} w_4 & 0 \\ 0 & -w_7 \end{bmatrix} .$$

5. OTHER ASPECTS OF THE THEORY

In Sections 2 and 3 the matrix W was allowed to be any positive real matrix, but in Section 4 only diagonal matrices W were considered. The reason for this restriction was that we considered only electrical network elements which could be placed in one edge. It is physically more realistic to allow mutual inductances between edges, as for example, if two edges correspond to the two windings of a transformer. In this case it is still true that W will be positive real, so that the theory applies and Z is again positive real.

If the requirement that W be Hermitian on the real axis is weakened to the requirement that W be almost right definite there, then still more network elements may be introduced. The resulting networks are called non-reciprocal (in contrast to the reciprocal networks we have studied). The basic non-reciprocal device is called the gyrator; it was introduced by Tellegen [19]. The use of gyrators is now standard in network theory textbooks.

When the requirement of finite dimensionality is dropped, the resulting network is called a Hilbert port [23]. These networks are under active study at the present time [2], [10], [11], [13], [24], [25].

Once it is known that impedance matrices are positive real, it becomes natural to treat the question of network synthesis. That is, can all positive real matrices be realized as impedance matrices of n-port networks? In answering this question, it is customary to allow the networks to contain ideal transformers; these transformers allow perfect coupling between edges with no inductance involved--thus they work for direct current ($\lambda = 0$). Using ideal transformers, the question was answered in the affirmative for 1-ports by Brune [8] and for n-ports by Bayard [4].

The synthesis question becomes much more difficult when only resistors, capacitors and inductors are allowed. For the 1-port case Bott and Duffin have shown that the answer is still affirmative [6]. For the n-port case, $n \geq 2$, the answer is different. It may be easily seen that the matrix

$$\begin{bmatrix} 1 & 2 \\ 2 & 5 \end{bmatrix}$$

cannot be synthesized without transformers; moreover there are no general conditions for determining when transformerless synthesis is possible. An interesting unsolved special case is to determine which real constant matrices can be realized by networks containing only resistors.

In this paper the properties of impedance matrices have been derived starting with descriptions of certain specific network elements. An alternative approach is to view the network as a "black box," whose time behavior is subject to certain basic physical principles, such as passivity, causality and translation invariance. It may be proved from these fundamental principles that the impedance matrix must be positive real [16].

REFERENCES

1 W. N. Anderson, Jr., R. J. Duffin and G. E. Trapp, Matrix operations induced by network connections. SIAM J. Control, 13(1975)446-461.

2 W. N. Anderson, Jr., and G. E. Trapp, Shortened operators II. SIAM J. Appl. Math., 28(1975)160-171.

3 H. Bart, M. A. Kaashoek and D. C. Lay, Relative inverses of finite meromorphic operator functions. Indag. Math., 36(1974)217-259.

4 M. Bayard, Synthèse des reseaux possifs à on nombre quelconque de paires de bornes connaisant leurs matrices d'impédances au d'admittances. Bull. Société Franciase Electriciens, 9(1949)497-502.

5 J. Bendat and S. Sherman, Monotone and convex operator functions. Trans. AMS, 79(1955)58-71.

6 R. Bott and R. J. Duffin, Impedance synthesis without use of transformers. J. Appl. Phys., 20(1949)816.

7 R. Bott and R. J. Duffin, On the algebra of networks. Trans. AMS, 74(1953)99-109.

8 O. Brune, Synthesis of a finite two terminal network whose driving point impedance is a prescribed function of frequency. J. Math and Phys., 10(1931)191-236.

9 I. Cederbaum, On equivalence of resistive n-port networks. IEEE Trans. Circuit Theory CT-12(1955)338-344.

10 V. Dolezal, Hilbert networks. SIAM J. Control, 12(1974)755-788.

11 V. Dolezal and A. H. Zemanian, Hilbert networks II--some qualitative properties. SIAM J.Control, 13(1975)153-161.

12 R. J. Duffin, Elementary operations which generate network matrices. Proc. AMS, 6(1955)335-339.

13 H. Flanders, Infinite networks I--resistive networks. IEEE Trans. Circuit Theory CT-18(1971)326-331.

14 D. Haxony, Elements of network synthesis. (New York: Reinhold, 1963).

15 K. Lowner, Uber monotone Matrixfunktionen. Math. Z. 38(1934) 177-216.

16 R. W. Newcomb, Linear multiport synthesis. (New York: McGraw-Hill, 1966).

17 P. I. Richards, A special class of functions with positive real part in the half-plane. Duke Math. J., 14(1947) 777-786.

18 F. Riesz and B. Sz-Nagy, Functional analysis. (New York: Ungar, 1955).

19 B. D. H. Tellegen, The gyrator, a new network element. Phillips Res. Repts., 3(1948)81-101.

20 L. Weinberg and P. Slepian, Positive real matrices. J. Math. Mech., 9(1960)71-84.

21 E. P. Wigner, Nuclear reactions and level widths. Amer. J. Phys., 17(1949)99-109.

22 E. P. Wigner, On a class of analytic functions from the quantum theory of collisions. Ann. Math. 53(1951)36-67.

23 A. H. Zemanian, The Hilbert port. SIAM J. Appl. Math., 18(1970) 98-138.

24 A. H. Zemanian, Realizability theory for continuous linear systems. (New York: Academic Press, 1972).

25 A. H. Zemanian, Infinite networks of positive operators. Circuit Theory & Appl., 2(1974)69-78.

George E. Trapp
Department of Statistics and Computer Science
University of West Virginia
Morgantown, West Virginia

and

William N. Anderson, Jr.
Department of Mathematics
East Tennessee State University
Johnson City, Tennessee 37601

This research was partially supported by NSF Grant No. MCS-75-07015.

R J DUFFIN AND T D MORLEY

Inequalities induced by network connections

ABSTRACT

It has been found that interesting mathematical relationships arise from a vectorial generalization of Kirchhoff's and Ohm's laws, in which the "resistors" become positive semi-definite (PSD) linear operators. In analogy to the parallel connection of resistors, Anderson and Duffin studied the parallel sum P:S of two PSD operators on a finite dimensional space, defined by $R:S = R(R+S)^{\dagger}S$. This paper extends the results of Anderson and Duffin that $||R:S|| \leq ||R|| : ||S||$ and $\mathrm{tr}[R:S] \leq \mathrm{tr}R:\mathrm{tr}S$ to a wide class of operations derived by a vectorial analog of Kirchhoff's and Ohm's laws. These inequalities remain true in Hilbert space.

1. INTRODUCTION

It has been found that interesting mathematical relationships develop from the analog of Kirchhoff's laws, in which current and voltage become vectors, rather than scalars, and the resistors become linear operators, Z_i. For example, Anderson and Duffin [2] developed the vectorial analog of series and parallel addition of resistors. In the parallel connection of the two resistors Z_1 and Z_2, the joint resistance R is given by the well-known formula

$$R^{-1} = Z_1^{-1} + Z_2^{-1} \quad . \tag{1}$$

This may be written in the form

$$R = Z_1(Z_1 + Z_2)^{-1}Z_2 \quad . \tag{2}$$

Anderson and Duffin used the latter formula to define "the parallel sum" of positive semi-definite operators on a finite dimensional space. It turns out that the joint resistance operator R is always defined if the inverse $(Z_1 + Z_2)^{-1}$ is interpreted as the Moore-Penrose pseudo-inverse.

Let the function on the right side of (2) be denoted by $\phi(.,.)$ when the arguments are scalars and by $\Phi(.,.)$ when the arguments are operators. Thus (2) may be written in the abbreviated form $R = \Phi(Z_1,Z_2)$. In this notation, Anderson found the following inequality for the norm of the parallel sum operator R

$$||\Phi(Z_1,Z_2)|| \leq \phi(||Z_1||,||Z_2||) \quad . \tag{3}$$

He also showed that the trace of R satisfies the similar inequality

$$tr[R] \leq \phi(tr[Z_1],tr[Z_2]) \quad . \tag{4}$$

In this paper we consider arbitrary networks with n branches. Let the branch resistances be $Z_1,Z_2,\ldots,Z_n$ and let R be the joint resistance of the particular connection under consideration. Then by applying Kirchhoff's laws in the usual way one determines

$$R = \phi(Z_1,Z_2,\ldots,Z_n) \tag{5}$$

where ϕ is a rational function termed the joint resistance function. If the scalar resistors are replaced by operators then a vectorial interpretation of Kirchhoff's laws leads to a formula for the joint resistance operator as

$$R = \Phi(Z_1,Z_2,\ldots,Z_n) \quad . \tag{6}$$

We prove that the norm of the joint resistance operator satisfies the inequality

$$||\Phi(Z_1,\ldots,Z_n)|| \leq \phi(||Z_1||,\ldots,||Z_n||) \quad . \tag{7}$$

A similar formula holds for trace. These inequalities hold even when the Z_i are operators on a Hilbert space.

To treat the problem posed, we find it convenient to express Kirchhoff's laws as a pair of matrix equations. The first equation gives Kirchhoff's current law, and the second equation gives Kirchhoff's voltage-drop law. These equations involve a matrix A, which is the node-arc incidence matrix of the network. However, all of the proofs hold for an arbitrary matrix. This observation shows that our theorems extend to general linear electromechanical systems.

Although the definition of these operator functions Φ was motivated by electrical network considerations, applications exist in the fields of statistics [7], Hilbert space theory [12], structural mechanics [11] and chemistry [14].

Another application in electrical work is the so-called nonreciprocal networks. These networks contain elements called gyrators. To treat this generalization, operators Z are introduced, and termed almost positive definite (see [2], [10], [14]) if

$$(Zx,x) \geq 0 \quad \text{for all } x \text{, and}$$

$$(Zx,x) = 0 \quad \text{only if } Zx = 0 \quad .$$

2. A VECTORIAL FORMULATION OF KIRCHHOFF'S LAWS

We consider the vectorial analogue of Kirchhoff's and Ohm's laws. The current and voltage become vectors, and the resistors become linear operators. Consider a directed graph G with m nodes and n arcs. We think of G as directions to connect together resistors. Let $x_1,\ldots,x_n$ be the currents in the arcs $1,2,\ldots,n$. Let the (node to ground) voltages in the nodes be $v_1,\ldots,v_m$. Each of the $x_1,\ldots,x_n$ and $v_1,\ldots,v_m$ is to be a vector belonging to some finite dimensional inner product space $\mathcal{U}$. The analogue of Ohm's law on each arc is

$$Z_i x_i = w_i \quad ,$$

where w_i is the voltage drop across arc i. Let c be the current of the battery. Here we have $x_i \in \mathcal{U}$, $i = 1,\ldots,n$, $w_i \in \mathcal{U}$, $i = 1,\ldots,m$ and for each $i = 1,\ldots,n$, Z_i is a linear operator on $\mathcal{U}$; and $c \in \mathcal{U}$.

By Kirchhoff's laws we have

$$\sum_{j=1}^{n} \alpha_{ij} x_j = \begin{cases} c \text{ if } i = 1, \\ -c \text{ if } i = 2, \\ 0 \text{ otherwise} \end{cases}$$

and

$$w_j = \sum_{i=1}^{n} \alpha_{ij} v_i$$

where $\overline{A} = \{\alpha_{ij}\}$ is the incidence matrix of the directed graph. The voltage across nodes 1 and 2 is then $v_1 - v_2$.

To express the above in a more convenient form, we need some notation. Let us denote by $\mathcal{U}^r$ the inner product space of all r-tuples of elements of $\mathcal{U}$, written columnwise. If $\underset{\sim}{x}$ and $\underset{\sim}{y}$ are in $\mathcal{U}^r$, we write:

$$\underset{\sim}{x} = \begin{bmatrix} x_1 \\ \cdot \\ \cdot \\ x_i \\ \cdot \\ \cdot \\ x_r \end{bmatrix} \quad \text{and} \quad \underset{\sim}{y} = \begin{bmatrix} y_1 \\ \cdot \\ \cdot \\ y_i \\ \cdot \\ \cdot \\ y_r \end{bmatrix},$$

and define $(\underset{\sim}{x},\underset{\sim}{y}) = \sum_{i=1}^{n} (x_i,y_i)$. If B is a linear operator on $\mathcal{U}$, we define a linear operator $\overline{A} \otimes B : \mathcal{U}^n \to \mathcal{U}^m$ by the matrix

$$\begin{bmatrix} \alpha_{11}B & \alpha_{12}B & \cdots\cdots & \alpha_{1n}B \\ \vdots & & & \vdots \\ \alpha_{m1}B & \alpha_{m2}B & \cdots\cdots & \alpha_{mn}B \end{bmatrix}$$

where $\overline{A} = \{\alpha_{ij}\}$.

If $Z_1,\ldots,Z_n$ are each linear operators on $\mathcal{U}$, we define a linear operator on $\mathcal{U}^n$ by the following matrix:

$$\operatorname{diag}(Z_1,\ldots,Z_n) = \begin{bmatrix} Z_1 & & & & \bigcirc \\ & \cdot & & & \\ & & Z_i & & \\ & & & \cdot & \\ \bigcirc & & & & Z_n \end{bmatrix}.$$

Set $\overline{P} = \begin{bmatrix} 1 \\ -1 \\ 0 \\ \vdots \\ 0 \end{bmatrix}$. Then we have $\underset{\sim}{x} = \begin{bmatrix} x_1 \\ \vdots \\ x_n \end{bmatrix} \in \mathcal{U}^n$, $\underset{\sim}{v} = \begin{bmatrix} v_1 \\ \vdots \\ v_m \end{bmatrix} \in \mathcal{U}^m$,

and $\operatorname{diag}(Z_1,\ldots,Z_n)$: $\mathcal{U}^n \to \mathcal{U}^n$. With these definitions the previous system can be restated in compact form as:

$$(\overline{A} \otimes I)\underset{\sim}{x} = (\overline{P} \otimes I)c \qquad \text{(KV)}$$

$$\operatorname{diag}(Z_1,\ldots,Z_n)\underset{\sim}{x} - (\overline{A} \otimes I)^T \underset{\sim}{v} = 0 \qquad \text{(KV')}$$

The voltage across 1 and 2 is then $(P \otimes I)^T \underset{\sim}{v}$.

As an example take $m = n = 2$, $\mathcal{U} = \mathbb{R}^2$, $\overline{A} = \begin{bmatrix} 1 & 1 \\ -1 & -1 \end{bmatrix}$,

and $\overline{P} = \begin{bmatrix} 1 \\ -1 \end{bmatrix}$. Then (KV) can be written as

$$\begin{bmatrix} 1 & 0 & 1 & 0 \\ 0 & 1 & 0 & 1 \\ -1 & 0 & -1 & 0 \\ 0 & -1 & 0 & -1 \end{bmatrix} \begin{bmatrix} \xi_1 \\ \xi_2 \\ \xi_3 \\ \xi_4 \end{bmatrix} = \begin{bmatrix} 1 & 0 \\ 0 & 1 \\ -1 & 0 \\ 0 & -1 \end{bmatrix} \begin{bmatrix} \gamma_1 \\ \gamma_2 \end{bmatrix}$$

where

$$c = \begin{bmatrix} \gamma_1 \\ \gamma_2 \end{bmatrix}, \qquad x = \begin{bmatrix} \xi_1 \\ \xi_2 \\ \xi_3 \\ \xi_4 \end{bmatrix},$$

so $x_1 = \begin{bmatrix} \xi_1 \\ \xi_2 \end{bmatrix}$ and $x_2 = \begin{bmatrix} \xi_3 \\ \xi_4 \end{bmatrix}$.

The above equations are vectorial generalization of the electrical connection shown in figure 1.

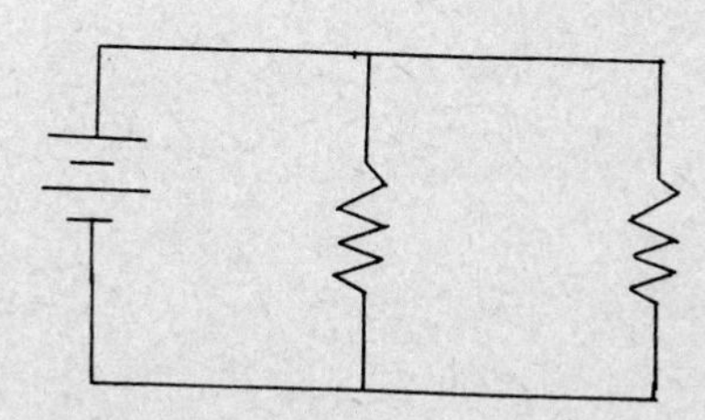

Figure 1: The Parallel Connection

Thus we see that Kirchhoff's and Ohm's laws are of the following form: given linear transformations $A : \mathcal{U} \to \mathcal{W}$, $Z : \mathcal{U} \to \mathcal{U}$, and $b \in$ range A,

find $\begin{bmatrix} x \\ v \end{bmatrix}$ such that

$$Ax = b, \qquad \text{(K)}$$

$$Zx - A^T v = 0 . \qquad \text{(K')}$$

Let us call a linear operator $Z: \mathcal{U} \to \mathcal{U}$ (where $\mathcal{U}$ is a finite dimensional inner product space) almost definite if $(Zx,x) = 0$ only when $Zx = 0$, if in addition $(Zx,x) \geq 0$, then Z is almost positive definite. It is well known that positive semi-**definite operators are almost definite.** (An operator is positive semi-definite if $(Zx,x) \geq 0$ and $Z = Z^T$, where Z^T is the adjoint of Z.) If Z is a matrix, then $\text{car}(Z) = \ker(Z)^{\perp}$, the orthogonal complement of the null space of Z, is called the carrier of Z.

The proofs of the following elementary facts may be found in [10]. The proofs are based on elementary applications of the Fredholm alternative for linear operators between finite dimensional inner product spaces.

Theorem 1 Given a linear operator $A: \mathcal{U} \to \mathcal{W}$ between two finite dimensional inner product spaces, and an almost definite linear operator $Z: \mathcal{U} \to \mathcal{U}$, there is a unique linear operator $\psi(Z)$ on range(A) satisfying the following properties:

(a) Given any $b \in$ range(A), there is an $x \in$ domain(A) and a $v \in \text{car}(A^T)$ such that

$$Ax = b \qquad \text{(K)}$$

$$Zx - A^T v = 0 . \qquad \text{(K')}$$

(b) $\psi(Z)b = v$

Theorem 2 Let Z be **almost** definite. Then so is $\psi(Z)$. If Z is almost positive (resp. negative) definite, then so is $\psi(Z)$, moreover $\psi(Z)^T = \psi(Z^T)$.

The $\psi(Z)$ constructed above is called the transfer resistance. The linear transformation A is called the structure operator, and Z is termed the constitutive operator.

We will occasionally use the abbreviation PSD for positive semi-definite.

A formula for $\psi(Z)$, involving the Moore-Penrose pseudo-inverse is given in [10]. Briefly let

$$Z = \begin{bmatrix} Z_{11} & Z_{12} \\ Z_{21} & Z_{22} \end{bmatrix} \quad \text{where} \quad Z_{11}: \text{car}(A) \to \text{car}(A),\ Z_{12}: \ker(A) \to \text{car}(A), \text{ etc.}$$

Then $\psi(Z) = A^{\dagger T}(Z_{11} - Z_{12}Z_{22}^{\dagger}Z_{21})A^{\dagger}$ restricted to range(A). The

operation Φ, termed the joint resistance, and defined later, will be of the form $P^T\psi(Z)P$ where A,Z,P, and b have a certain specialized structure.

3. POSITIVE SEMI-DEFINITE CONSTITUTIVE OPERATOR AND QUADRATIC PROGRAMMING

In his famous treatise Maxwell showed that one of Kirchhoff's laws may be replaced by a minimum principle. We develop an analogous approach in this section. We specialize system (K-K'),

$$Ax = b \qquad \text{(K)}$$
$$Zx - A^T v = 0 \qquad \text{(K')}$$

to the case where the operator Z is positive semi-definite, i.e., $(Zx,x) \geq 0$ for all x, and $Z = Z^T$. This allows one to replace (K-K') by a certain structured quadratic program. The structure of this program then allows one to draw various conclusions about the transfer resistance $\psi(Z)$. The physical significance of the specialization to positive semi-definite (PSD) linear operators Z is the specialization to gyrator free systems. As before all operators considered will be linear operators between finite dimensional inner product spaces.

Theorem 1 If Z is positive semi-definite, then so is $\psi(Z)$.

Proof See [10]. □

Theorem 2 (Power Inequality) Suppose the constitutive operator Z is positive semi-definite. Let $\hat{x}$ and b satisfy $A\hat{x} = b$ where A is the structure operator then:

$$(\psi(Z)b,b) \leq (Z\hat{x},\hat{x}) \qquad \text{(Power Inequality)}$$

Proof Consider the following quadratic program

Program (KQ): Find $VP(b) = \text{minimum } (Zx,x)$
subject to: $Ax = b$.

Here $VP(b)$ is termed the value of the program as a function of b. Since Z is positive semi-definite, and $b \in \text{range}(A)$, there is an x^* such that $(Zx^*,x^*) = VP$. Consider a variation $\delta x \in \ker(A)$. Then $0 = \delta(Zx^*,x^*) = 2(Zx^*,\delta x)$. Thus Zx^* is orthogonal to $\ker(A)$. Thus $Zx^* \in \text{range}(A^T)$. Therefore there is v such that $Zx^* = A^T v$. Combining this with $Ax^* = b$, we conclude that $\begin{bmatrix} x^* \\ v \end{bmatrix}$ is a solution to (K-K'). By Theorem 1 of Section 2 we conclude $(\psi(Z)b,b) = (Zx^*,x^*) \leq (Zx,x)$. □

The above proof leads us to the following characterization of the transfer resistance.

Theorem 3 $\psi(Z)$ is the unique symmetric operator such that $(\psi(Z)b,b) = VP(b)$ for all $b \in \text{range}(A)$. $VP(b)$ is defined as the solution of Program (KQ).

Proof Follows from the proof of the preceding theorem and Theorem 1 of Section **2.** □

The following two results illustrate the power of (KQ) in proving results on $\psi(Z)$. Recall that for PSD operators Y and Z, that $Y \leq Z$ means $Z - Y$ is positive semi-definite. This partial ordering is equivalent to the condition that $(Yx,x) \leq (Zx,x)$ for all x.

Theorem 4 $\psi(Z)$ is a monotone function of Z, in the following sense: If Y and Z are two positive semi-definite operators, and $Y \leq Z$, then $\psi(Y) \leq \psi(Z)$. The partial order is defined as above.

Proof Let $b \in \text{range}(A)$. Then

$$
\begin{aligned}
(\psi(Y)b,b) &= VP(b) = \min_{Ax=b} (Yx,x) \\
&\leq \min_{Ax=b} (Zx,x) \quad (\text{Since } Y \leq Z) \\
&= (\psi(Z)b,b). \quad \square
\end{aligned}
$$

The following is a type of global Lipschitz condition.

Theorem 5 Let Y and Z be positive semi-definite. Then there exists an operator F depending only on Z such that $0 \leq \psi(Y+Z) - \psi(Z) \leq F^T YF$.

Proof The first inequality follows from the previous theorem and the obvious relation $Y + Z \geq Z$. For the second inequality let F be an operator such that for any $b \in \text{range}(A)$, $(\psi(Z)b,b) = (ZFb,Fb)$. For example, we can define F by the relation $\bar{x} = Fb$ where $\bar{x}$ satisfies (K-K'). Then

$$AFb = 0$$

$$ZFb - A^T\psi(Z)b = 0$$

is seen to follow. Then consider

$$(\psi(X+Z)b,b) = \min_{Ax=b} ((Y+Z)x,x)$$

$$\leq ((Y+Z)\bar{x},\bar{x}) \quad \text{where} \quad \bar{x} = Fb$$

$$= (Y\bar{x},\bar{x}) + (Z\bar{x},\bar{x}) = (Y\bar{x},\bar{x}) + (\psi(X)b,b)$$

$$= (F^TYFb,b) + (\psi(X)b,b). \quad \square$$

Corollary 1 Let F be any operator satisfying $AFx = x$ for all $x \in \text{range}(A)$. (F is a generalized inverse of A.) Let Y be positive semi-definite. Then $0 \leq \psi(Y) \leq F^TYF$.

Proof Take $Z = 0$ in the preceding. $\square$

Corollary 2 Let Z be positive semi-definite. Let $\{E_n\}$ be a sequence of positive semi-definite operators such that $E_n \to 0$ and $n \to \infty$. Then $\psi(Z+E_n) \to \psi(Z)$ as $n \to \infty$.

Proof By Theorem 5 there is a matrix F such that $||\psi(Z+E_n) - \psi(Z)|| \leq ||F^TE_nF|| \leq ||F||^2||E_n||$. Since $||E_n|| \to 0$ as $n \to \infty$, the result follows. $\square$

In the next several sections we shall continually use the structure of (KQ) to obtain results.

Theorem 5 can be restated in the following manner: the function $Y \to F^TYF$ is a subgradient of the operator valued concave function $\psi(Z)$. Real valued concave (or convex) functions always have subgradients. Characterizing the class of Banach lattice valued concave (or convex) functions which possess subgradients is an open question.

4. SERIES PARALLEL DUALITY

In [2] Anderson introduced the following operation. Let S and T be positive semi-definite linear operators on a finite dimensional inner product space. Then we define $R:S = R(R+S)^{\dagger}S$, and call $R:S$ the parallel sum of R and S, see Figure 1. Here "dagger" denotes the Moore-Penrose pseudo-inverse.

The following results are needed about $R:S$. All are proved in [2]. However, (a) through (e) follows from the results of Sections 2 and 3.

<u>Lemma 1</u> Let R, S and T be positive semi-definite operators on a finite dimensional inner product space. Then

(a) $R:S$ is positive semi-definite

(b) $R:S = S:R$

(c) $(\alpha R) : (\alpha S) = \alpha(R:S)$ for $\alpha \geq 0$

(d) $R:(S:T) = (R:S):T$

(e) $(R:Sb,b) = \min\limits_{x+y=b} (Rx,x) + (Sy,y)$

(f) Let $R,S.\ T,U$ be positive semi-definite then
$(R + S):(T+U) \geq (R:T) + (S:U)$ where as before $Y \geq Z$ means $Y - Z$ is positive semi-definite.

(g) For $\alpha,\beta > 0$ define $\alpha:\beta = \alpha(\alpha+\beta)^{\dagger}\beta$ where $\gamma^{\dagger} = \gamma^{-1}$ if $\gamma \neq 0$, and $0^{\dagger} = 0$, then $||R:S|| \leq ||R||:||S||$.

(h) $tr[R:S] \leq tr[R]:tr[S]$.

We now obtain two theorems, one involving "+" and the other ":". These theorems are in some sense dual.

<u>Lemma 2</u> Let Y and Z be positive semi-definite. Then

$$\psi(Y+Z) \geq \psi(Y) + \psi(Z).$$

Combining this with $\psi(\alpha Z) = \alpha\psi(Z)$ for $\alpha \geq 0$, we conclude that ψ is concave.

<u>Proof</u>
$$\begin{aligned}
((\psi(Y) + \psi(Z))b,b) &= (\psi(Y)b,b) + (\psi(Z)b,b) \\
&= \min_{Ay=b} (Yy,y) + \min_{Ax=b} (Zz,z) \\
&\leq \min_{Ax=b} ((Y+Z)x,x) \\
&= (\psi(Y+Z)b,b). \quad \square
\end{aligned}$$

Theorem 1 Let Y and Z be positive semi-definite. Let $P:\mathcal{Y}\to\mathcal{X}$ be a linear transformation, where $\mathcal{X} = \text{range}(A)$, then

$$P^T\psi(Y + Z)P \geq P^T\psi(Y)P + P^T\psi(Z)P.$$

Proof Immediate. □

Theorem 2 Let Y and Z be positive semi-definite. Let $P:\mathcal{Y}\to\mathcal{X}$ be a linear transformation, where $\mathcal{X} = \text{range}(A)$, then

$$P^T\psi(Y{:}Z)P \leq (P^T\psi(Y)P) : (P^T\psi(Z)P).$$

Proof By Lemma 1(e) we have

$$(P^T\psi(Y)P : P^T\psi(Z)Pb,b) = \min_{c+d=b} (\psi(Y)Pc,Pc) + (\psi(Z)Pd,Pd) .$$ By

Theorem 3 of Section 4, the above is equal to

$$\min_{c+d=b} \min_{Ay=Pc} \min_{Az=Pd} (Yy,y) + (Zz,z).$$

If the constraints are related, then the minimum will decrease. Therefore,

$$\begin{aligned}(P^T\psi(Y)P : P^T\psi(z)Pb,b) &\geq \min_{Ax=Pb} \min_{y+z=x} (Yy,y) + (Zz,z)\\ &= \min_{Ax=Pb} ((Y{:}Z)x,x)\\ &= (P^T\psi(Y{:}Z)Pb,b) .\end{aligned}$$ □

It should be noticed that several of the above results are reformulations of results of Anderson and Trapp [4]. The present proofs, however, are simpler.

5. INEQUALITIES INDUCED BY NETWORK CONNECTIONS

We now specialize the system (K-K')

$$Ax = b \qquad (K)$$

$$Zx - A^Tv = 0 \qquad (K')$$

to the system (KV-KV')

$$(A \otimes I)\,\tilde{x} = (P \otimes I)c \qquad (KV)$$

$$\text{diag}(Z_1,\ldots Z_n)\tilde{x} - (A \otimes I)^T\tilde{v} = 0 \qquad (KV')$$

where

$\overline{A}$	an m by n matrix
$\overline{P}$	an m by 1 matrix (column vector) with $\overline{P} \in \text{range}(\overline{A})$
$Z_1,\ldots,Z_n$	linear operators on some finite dimensional inner product space V.

System (KV-KV') reduces to (K-K') if we set

$$A = (\overline{A} \otimes I) \tag{1a}$$

$$b = (\overline{P} \otimes I)c \tag{1b}$$

$$Z = \text{diag}(Z_1,\ldots,Z_n) \tag{1c}$$

<u>Theorem 1</u> Suppose $Z_1,\ldots,Z_n$ are **al**most positive definite. Then there exists a unique linear operator $\Phi(Z_1,\ldots,Z_m) : V \to V$ such that for any $c \in V$

$$\Phi(Z_1,\ldots,Z_n)c = (P \otimes I)^T \underset{\sim}{v}$$

where $\begin{bmatrix} \underset{\sim}{x} \\ \underset{\sim}{v} \end{bmatrix}$ is any solution of (KV-KV').

<u>Proof</u> If $Z_1,\ldots,Z_n$ are almost positive definite operators on V, then $Z = \text{diag}(Z_1,\ldots,Z_n)$ is almost positive definite operator on V^n. Since $\overline{P} \in \text{range}(\overline{A})$, we have $(P \otimes I)c \in \text{range}(A \otimes I)$. Thus, by Theorem 1 of Section 2, we have a solution of (KV-KV'). Set $\Phi(Z_1,\ldots,Z_n)c = [P \quad I]^T \underset{\sim}{v}$. By the linearity of (KV-KV'), $\Phi(Z_1,\ldots,Z_n)$ is clearly a linear operator. By Theorem 1 of Section 2, if $\begin{bmatrix} \underset{\sim}{x} \\ \underset{\sim}{v} \end{bmatrix}$ and $\begin{bmatrix} \underset{\sim}{x}' \\ \underset{\sim}{v}' \end{bmatrix}$ both solve (KV-KV'), then $(A \otimes I)^T v = (A \otimes I)^T v'$. Since $P \in \text{range}(A)$, we have $\ker(A \otimes I)^T \subseteq \ker(P \otimes I)^T$, and thus $(P \otimes I)^T \underset{\sim}{v} = (P \otimes I)^T \underset{\sim}{v}'$. □

We term $\Phi(Z_1,\ldots,Z_n)$ the joint resistance. In terms of the transfer resistance (with A and Z as defined in (1)) we have

$$\Phi(Z_1,\ldots,Z_n) = (\overline{P} \otimes I)^T \psi(Z)\ (\overline{P} \otimes I).$$

As an example of Φ we have parallel addition [2], the analogue of the connection in Figure 1.

Proposition 1 Let Z_1 and Z_2 be almost positive definite linear operators on some finite dimensional vector space.

Let $\bar{A} = \begin{bmatrix} 1 & 1 \\ -1 & -1 \end{bmatrix}$ and $\bar{P} = \begin{bmatrix} 1 \\ -1 \end{bmatrix}$ and $Z_1 : Z_2 = Z_1(Z_1 + Z_2)^{\dagger} Z_2$.

Then $\Phi(Z_1, Z_2) = Z_1 : Z_2$.

Proof See [2]. □

The following results follow immediately from the corresponding results in Sections 2 and 4, on the transfer resistance, ψ. Recall that an almost positive definite operator Z satisfies $(Zx, x) \geq 0$, with equality only when $Zx = 0$.

Theorem 2 Let $Z_1, \ldots, Z_n$ be almost positive definite operators on V and let $\Phi(Z_1, \ldots, Z_n)$ be the joint resistance.
Then

(a) $\Phi(Z_1, \ldots, Z_n)$ is almost positive definite operator on V.

(b) Let $\alpha \geq 0$. Then $\Phi(\alpha Z_1, \ldots, \alpha Z_n) = \alpha \Phi(Z_1, \ldots, Z_n)$.

(c) $\Phi(Z_1, \ldots, Z_n) = \lim_{\varepsilon \to 0^+} \Phi(Z_1 + \varepsilon I, \ldots, Z_n + \varepsilon I)$.

(d) Suppose $\Phi(Z_1, \ldots, Z_n)c = z$. Then there is an $\underset{\sim}{x}$ and a $\underset{\sim}{v}$ such that $x \in V^n$, $v \in V^m$

$$(\bar{A} \otimes I)x = (\bar{P} \otimes I)c$$

$$\text{diag}(Z_1, \ldots, Z_n)\underset{\sim}{x} - (A \otimes I)^T \underset{\sim}{v} = 0$$

and $z = (\bar{P} \otimes I)^T \underset{\sim}{v}$.

(e) If $Z_1, \ldots, Z_n$ are positive semi-definite, then so is $\Phi(Z_1, \ldots, Z_n)$. Moreover in this case $(\Phi(Z_1, \ldots, Z_n)c, c) = \min_{S(c)} (Z\underset{\sim}{x}, \underset{\sim}{x})$ where $Z = \text{diag}(Z_1, \ldots, Z_n)$ and $S(c) = \{\underset{\sim}{x} : (\bar{A} \otimes I)x = (\bar{P} \otimes I)c\}$.

(f) If $Y_1, \ldots, Y_n$, $Z_1, \ldots, Z_n$ are positive semi-definite, and $Y_i \leq Z_i$, $i = 1, \ldots, n$, then $\Phi(Y_1, \ldots, Y_m) \leq \Phi(Z_1, \ldots, Z_m)$.

Associated with any $\overline{A}$ and $\overline{P}$ is an ordinary scalar joint resistance.

Theorem 3 Suppose $\lambda_1,\ldots,\lambda_m \geq 0$. Then there is a unique non-negative real number $\phi(\lambda_1,\ldots,\lambda_n)$ such that for any solution

$$\begin{bmatrix} x \\ v \end{bmatrix}, \quad x \in \mathbb{R}^n, \quad v \in \mathbb{R}^m, \quad \text{of}$$

$$Ax = \overline{P}$$

$$\operatorname{diag}(\lambda_1,\ldots,\lambda_n)x - \overline{A}^T v = 0$$

we have $\phi(\lambda_1,\ldots,\lambda_n) = \overline{P}^T v$.

Proof Apply Theorem 1 to the case $V = \mathbb{R}$. $\square$

The real number $\phi(\lambda_1,\ldots,\lambda_n)$ is the ordinary joint resistance when resistors $\lambda_1,\ldots,\lambda_n$ are connected together. See [15]. Many simple methods for the calculation of ϕ exist. Methods for calculating Φ are more complex. Thus it is of interest to find interrelations between Φ and ϕ.

Theorem 4 Let $\lambda_1,\ldots,\lambda_n \geq 0$. Let $I:V \to V$ be the identity operator. Then

$$\Phi(\lambda_1 I,\ldots,\lambda_m I) = \phi(\lambda_1,\ldots,\lambda_m)I.$$

Proof Let $\phi(\lambda_1,\ldots,\lambda_m) = \alpha$. Then by Theorem 3 there is an x and a v

$$x = \begin{bmatrix} \chi_1 \\ \vdots \\ \chi_m \end{bmatrix} \qquad v = \begin{bmatrix} \nu_1 \\ \vdots \\ \nu_n \end{bmatrix}$$

with $\overline{A}v = \overline{P}$ and $\operatorname{diag}(\lambda_1,\ldots,\lambda_n)x - \overline{A}^T v = 0$. Given $c \in V$ set

$$\underset{\sim}{x}' = \begin{bmatrix} \chi_1 c \\ \vdots \\ \chi_n c \end{bmatrix} = x \otimes c \quad \text{and} \quad \underset{\sim}{v}' = \begin{bmatrix} \nu_1 c \\ \vdots \\ \nu_1 c \end{bmatrix} = v \otimes c \quad .$$

Then

$$(\overline{A} \otimes I)x' = (\overline{A} \otimes I)(x \otimes \mathbf{c}) = (\overline{A}x) \otimes c = \overline{P} \otimes c = (\overline{P} \otimes I)c.$$

(Here we have used the well-known fact that $(B \otimes C)(D \otimes E) = (BC \otimes CE)$. Also

$$\mathrm{diag}(\lambda_1 I,\dots,\lambda_n I)\underset{\sim}{x}' = (\mathrm{diag}(\lambda_1,\dots,\lambda_n)\otimes I)\underset{\sim}{x}'$$

$$= (A^T x) \otimes c = (A \otimes I)^T (x \otimes c).$$

Thus $\underset{\sim}{x}'$ and $\underset{\sim}{v}'$ solve (KV-KV'). The Theorem follows. □

The following theorem is the first of several that relate values of Φ and ϕ, for general positive semi-definite $Z_1,\dots,Z_n$.

Theorem 5 Let $Z_1,\dots,Z_n$ be positive semi-definite. Then the joint resistance Φ satisfies:

$$||\Phi(Z_1,\dots,Z_n)|| \leq \phi(||Z_1||,\dots,||Z_n||).$$

Proof Since $Z_i \leq ||Z_i||I$, we have

$$\Phi(Z_1,\dots,Z_n) \leq \Phi(||Z_1||I,\dots,||Z_n||I)$$

$$= \phi(||Z_1||,\dots,||Z_n||)I.$$

Taking norms, the result follows. □

Theorem 5 is a generalization of $||Z_1 : Z_2|| \leq ||Z_1||:||Z_2||$. The following provides a generalization of

$$((Z_1 : Z_2)c,c) \leq (Z_1 c,c) : (Z_2 c,c) \quad .$$

Theorem 6 Let $Z_1,\dots,Z_n$ be positive semi-definite. Then

$$(\Phi(Z_1,\dots,Z_n)c,c) \leq \phi((Z_1 c,c),\dots,(Z_n c,c)) \quad .$$

Proof By Theorem 2(e) we have

$$(\Phi(Z_1,\dots,Z_n)c,c) = \min_{S(c)} (Z\underset{\sim}{x},\underset{\sim}{x}) \tag{1}$$

where $S(c) = \{\underset{\sim}{x}:(\bar{A}\otimes I)\underset{\sim}{x} = (\bar{P}\otimes I)c\}$. Setting $S'(c) = \{\underset{\sim}{x} \in S(c) :$ $\underset{\sim}{x} = y \otimes c \}$, (Recall that $\otimes$ denotes kornecker product.) We now have $S'(c) \subseteq S(c)$, so

$$(\Phi(Z_1,\dots,Z_n)c,c) \leq \min_{S'(c)} (Z\underset{\sim}{x},\underset{\sim}{x}) \quad \text{, where}$$

$$S'(c) = \{(y \otimes c) : (\bar{A} \otimes I)(y \otimes c) = (\bar{P} \otimes I)c\} \quad .$$

But $(\bar{A} \otimes I)(y \otimes c) = \bar{A}y \otimes c$, so $S'(c) = \{(y \otimes c) : \bar{A}y = \bar{P}c\}$,

and we have

$$\min_{S'(c)} (Z(y \otimes c), y \otimes c) = \min_{\bar{A}y=\bar{P}} \sum \eta_i^2 (Z_i c, c), \quad y = \begin{bmatrix} \eta_1 \\ \vdots \\ \eta_n \end{bmatrix} ,$$

$$= \phi((Z_1 c, c), \ldots, (Z_n c, c)) \quad .$$

by Theorem 2(e) applied to ϕ. □

Before we can prove the analogue of Theorem 5 for trace instead of norm, we need the following result.

<u>Theorem 7</u> Let $Y_1, \ldots, Y_n$ and $Z_1, \ldots, Z_n$ be positive semi-definite linear operators on a finite dimensional inner product space. Then

(a) $\Phi(Y_1 + Z_1, \ldots, Y_n + Z_n) \geq \Phi(Y_1, \ldots, Y_n) + \Phi(Z_1, \ldots, Z_n)$

and

(b) $\Phi(Y_1 : Z_1, \ldots, Y_n : Z_n) \leq \Phi(Y_1, \ldots, Y_n) : \Phi(Z_1, \ldots, Z_n)$.

<u>Proof</u> The result follows from Theorems 4.1 and 4.2. □

Suppose $Z_1, \ldots, Z_r$ are positive semi-definite. Then since parallel addition is associative (see [2]) we may define

$$\prod_{j=1}^{r} : Z_i = Z_1 : Z_2 : \ldots : Z_r \quad .$$

<u>Corollary 1</u> Let Z_{ij}, $i = 1, \ldots, r$, $j = 1, \ldots, r$, all be positive semi-definite linear operators on a finite dimensional inner product space. Then

(a) $\Phi(\sum_{j=1}^{r} Z_{ij}, \ldots, \sum_{j=1}^{r} Z_{nj}) \geq \sum_{j=1}^{r} \Phi(Z_{1j}, \ldots, Z_{nj})$

and

(b) $$\Phi\left(\prod_{j=1}^{r} : Z_{1j}, \ldots, \prod_{j=1}^{r} : Z_{nj}\right) \leq \prod_{j=1}^{r} : \Phi(Z_{1j}, \ldots, Z_{nj}) .$$

Proof Apply induction to the previous result. □

We can now prove the analogue of Theorem 5 for trace instead of norm. This generalizes $tr[R:S] \leq tr[R] : tr[S]$ proved for PSD R and S in [2]. For diagonal PSD R and S, $tr[R:S] \leq tr[R]:[S]$ is Minkowski's inequality.

Theorem 8 Let $Z_1, \ldots, Z_n$ be positive semi-definite operators on V. Then the joint resistance, Φ, satisfies

$$tr[\Phi(Z_1, \ldots, Z_n)] \leq \phi(tr[Z_1], \ldots, tr[Z_n]).$$

Proof Let $\{e_i\}_{i=1}^{r}$ be a basis of V. Then setting $R = \Phi(Z_1, \ldots, Z_n)$ we have

$$tr[R] = \sum_{j=1}^{r} (Re_j, e_j) \leq \sum_{j=1}^{r} \phi((Z_1 e_j, e_j), \ldots, (Z_n e_j, e_j))$$

by Theorem 6. Now applying Corollary 1(a) we conclude

$$tr[R] \leq \phi\left(\sum (Z_1 e_j, e_j), \ldots, \sum (Z_n e_j, e_j)\right)$$

$$= \phi(tr[Z_1], \ldots, tr[Z_n]). \quad \square$$

The above theorems hold for an essentially arbitrary $\overline{A}$ and $\overline{P}$. It is interesting, however, to look at special cases.

Proposition 2 Let $\overline{A}$ be the node-arc incidence matrix of the graph in Figure 2,

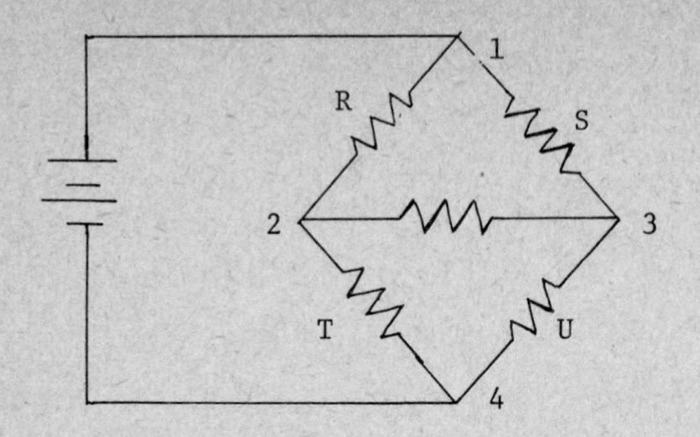

Figure 2: A Wheatstone Bridge

and set $\overline{P} = \begin{bmatrix} 1 \\ 0 \\ 0 \\ -1 \end{bmatrix}$. Then writing $Z_1 = R$, $Z_2 = S$, $Z_3 = T$,

$Z_4 = U$, $Z_5 = V$ we have

$$\Phi(R,S,T,U,V) = W(R,S,T,U,V)$$

$$= (R-T) + R(R+S+V)^{\dagger}R + R(R+S+V)^{\dagger}V(L+T+U)^{\dagger}(T+M)$$

$$- \; T(L+T+U)^{\dagger}(T+M)$$

where $L = V - V(R+S+V)^{\dagger}V$, and $M = V(R+S+V)^{\dagger}R$. The scalar joint resistance can be written as

$$\phi(r,s,t,u,v) = w(r,s,t,u,v)$$

$$= \frac{rst + rsu + rtv + rtu + stv + tuv}{rt + ru + rv + st + su + sv + tv + uv} .$$

These formulas may be verified by computation. The proof is omitted. An immediate consequence is

Corollary 2 Let $W(R,S,T,U,V)$ and $w(r,s,t,u,v)$ be as above. Then

(a) $||W(R,S,T,U,V)|| \leq w(||R||,||S||,||T||,||U||,||V||)$

(b) $(W(R,S,T,U,V)b,b) \leq w((Rb,b), (Sb,b),(Tb,b),(Ub,b),(Vb,b))$

(c) $tr[W(R,S,T,U,V)] \leq w(tr[R],tr[S],tr[T],tr[U],tr[V])$.

Even though the exact formulas for Φ, and ϕ are computed for the Wheatstone bridge above, the results of Corollary 2 are not at all obvious from the algebraic formulas. This, then, illustrates the power of the general theory.

6. GENERALIZATIONS TO HILBERT SPACE

In [12] Fillmore and Williams defined an operator

$$Z_1 : Z_2 = \sqrt{Z_1}\,(\sqrt{Z_1 + Z_2}^{\dagger}\,\sqrt{Z_1})^{*}\,(\sqrt{Z_1 + Z_2}^{\dagger}\,\sqrt{Z_2})\,\sqrt{Z_2} \tag{1}$$

for bounded, positive semi-definite operators on a Hilbert space. They showed that for positive semi-definite Z_1 and Z_2 one has

$\text{range}(\sqrt{Z_1}) \subseteq \text{range}(\sqrt{Z_1 + Z_2})$ and therefore the pseudo-inverses in (1) are well defined and bounded. They proved several properties of this operation that were analogues of the parallel addition on finite dimensional inner product spaces. (see Section 5). However, they did not show associativity, i.e.,

$$Z_1 : (Z_2 : Z_3) = (Z_1 : Z_2) : Z_3 \quad .$$

In [5] Anderson and Trapp used an alternate but equivalent definition, based on the shorted operator of Anderson and Krein. They proved associativity, however they did not show that

$$\text{tr}[Z_1 : Z_2] \leq \text{tr}[Z_1] : \text{tr}[Z_2].$$

(For positive semi-definite operators, trace is always defined, although it may be at $+\infty$.)

To prove generalizations of the above we need the use of the shorted operator.

Theorem 1 ([5] or [13]) Let $\mathcal{U}$ be a Hilbert space and let $\mathcal{S}$ be a closed subspace. Let Z be a positive semi-definite (bounded) linear operator on $\mathcal{U}$. Then there exists a unique symmetric linear operator $\mathcal{S}(Z) : \mathcal{S} \to \mathcal{S}$ such that

$$(\mathcal{S}(Z)s, s) = \inf_{t \in \mathcal{S}} (Z(s+t), (s+t)) \quad \text{for all} \quad s \in \mathcal{S} \quad .$$

The straightforward generalization of (K-K') to Hilbert space need not have a solution even if A has a closed range. We replace it with

<u>Theorem 2</u> Given $A: \mathcal{U} \to \mathcal{W}$ with closed range and $Z: \mathcal{U} \to \mathcal{U}$ a positive semi-definite linear operator on $\mathcal{U}$, there exists a unique $\mathcal{S}(Z)$: range(A) $\to$ range(A) such that

$$(\mathcal{S}(Z)b,b) = \inf_{Ax=b} (Zx,x). \qquad \text{(K H)}$$

<u>Proof</u> Let $\mathcal{S} = \text{car}(A)$ and set $\psi(Z) = A^{\dagger T}\mathcal{S}(Z)A^{\dagger}$ restricted to range(A). Then

$$\begin{aligned}(\psi(Z)b,b) &= (A^{\dagger T}\mathcal{S}(Z)A^{\dagger}b,b)\\ &= (\mathcal{S}(Z)A^{\dagger}b,A^{\dagger}b)\\ &= \inf_{x\in \ker(A)} (Z((A^{\dagger}b)+x),A^{\dagger}b+x)\\ &= \inf_{Ax=b} (Zx,x) \quad . \quad \square\end{aligned}$$

Letting $\overline{A} : \mathbb{R}^n \to \mathbb{R}^m$ be a matrix, and letting $\overline{P} \in$ range(A), then by setting $A = (\overline{A} \otimes I)$, where I is the identity on a Hilbert space, we may define

$$\Phi(Z_1,\ldots,Z_n) = (\overline{P} \otimes I)^T\mathcal{S}(Z)\ (\overline{P} \otimes I)$$

with $Z = \text{diag}(Z_1,\ldots,Z_n)$ and $\mathcal{S}$ as above. We then have the following inequalities for PSD operators on a Hilbert space

$$||\Phi(Z_1,\ldots,Z_n)|| < \phi(||Z_1||,\ldots,||Z_n||)$$

$$(\Phi(Z_1,\ldots,Z_n)b,b) \le \phi((Z_1b,b),\ldots,(Z_nb,b))$$

$$\text{tr}[\Phi(Z_1,\ldots,Z_n)] \le \phi(\text{tr}[Z_1],\ldots,\text{tr}[Z_n]) \quad .$$

In fact, with the above definition, the following theorems (if Z (or $Z_1,\ldots,Z_n$) are assumed to be positive semi-definite) hold with minor modification in their proofs: Theorems 3.1, 3.2, 3.4, Corollary 3.1, Theorems 4.1, 4.2, 5.4, 5.5, 5.6, 5.7, 5.8. The limiting process needed in the proof of the analogue of Theorem 5.8 poses no problem because the functions considered are monotone.

Letting $Z_1 : Z_2$ be defined as in (1), Anderson and Trapp [5] have shown that

$$(Z_1 : Z_2 x, x) = \inf_{y+z=x} (Z_1 y, y) + (Z_2 z, z) \quad .$$

Thus parallel sum is a special case of the operator Φ with

$\overline{A} = \begin{bmatrix} 1 & 1 \\ -1 & -1 \end{bmatrix}$ and $\overline{P} = \begin{bmatrix} 1 \\ -1 \end{bmatrix}$, and the theorems of Section 6 apply.

If the range of A is not closed, but the null space is closed, then $\psi(Z)$ is still defined although it may be unbounded.

7. GENERALIZATIONS

R. J. Duffin and G. E. Trapp in analogy with the hybrid electrical connection, studied the hybrid sum $R*S$, of two

partioned n by n PSD matrices $R = \begin{bmatrix} R_{11} & R_{12} \\ R_{21} & R_{22} \end{bmatrix}$ and

$S = \begin{bmatrix} S_{11} & S_{12} \\ S_{21} & S_{22} \end{bmatrix}$ where R_{11}, R_{22}, S_{11}, and S_{22} are square, with

the same partition being used in both R and S. Defining

$|||A||| = \begin{bmatrix} ||A_{11}|| & 0 \\ 0 & ||A_{22}|| \end{bmatrix}$ for matrices of the same size and same

partition as R, they showed that $|||R*S||| \leq |||R|||*|||S|||$.
They also showed a similar formula for a type of generalized trace. Generalizations of the theorems in this paper will prove analogous results for a wide variety of "hybrid type" electrical connections, see [17].

REFERENCES

1 W. N. Anderson, Jr., Series and parallel addition of operators, Doctoral Dissertation, Carnegie-Mellon Univ., 1968.

2 W. N. Anderson, Jr., and R. J. Duffin, Series and Parallel addition of matrices. J. Math. Anal. Appl. 26(1969)576-594.

3 W. N. Anderson, Jr., Shorted operators. SIAM J. Appl. Math. 20(1971) 520-525.

4 W. N. Anderson, Jr., R. J. Duffin and G. E. Trapp, Matrix operations induced by network connections. SIAM J. Control 13(1975)446-461.

5 W. N. Anderson, Jr., and G. E. Trapp, Shorted operators II. SIAM J. Appl. Math. 28(1975)61-71.

6 W. N. Anderson, Jr., and G. E. Trapp, Matrix operations induced by network connections II.Bull.Calc.Math., 69(1977)319-330.

7 W. N. Anderson, Jr., G. D. Kleindorfer, P. R. Kleindorfer and M. B. Woodroofe, Consistent estimates of the parameters of a linear system. Ann. Math. Statis. 40(1969) 1064-1075.

8 R. J. Duffin and G. E. Trapp, Hybrid addition of matrices-a network theory concept. Applicable Anal. 2(1972)241-254.

9 R. J. Duffin, Network models. SIAM-AMS Proc. 3(1971)65-91.

10 R. J. Duffin and T. D. Morley, Almost definite operators and electromechanical systems. SIAM J. Appl. Math. 35(1978) 21-30.

11 S. J. Fenves, Structural analysis by networks, matrices and computers. J. Struc. Div., Proc. ASCE 92(1966) 199-221.

12 P. A. Fillmore and J. P. Williams, On operator ranges. Adv. in Math., 7(1971)254-281.

13 M. A. Klein, The theory of self-adjoint extensions of semi-bounded Hermitian transformations and its applications. Math. Coll. Moscow, 20(1974)431-495, 21(1947)365-404.

14 A. Predelson and G. Oster, Chemical reaction dynamics, part II - reaction networks. Arch. Rational Mech. 57(1974) 31-98.

15 S. Seshu and M. Reed, Linear graphs and electrical networks, (Reading, Mass.: Addison Wesley, 1961).

16 G. E. Trapp, Algebraic operations derived from electrical networks. Doctoral Dissertation, Carnegie-Mellon Univ. (1970).

17 R. J. Duffin and T. D. Morley, Inequalities induced by electrical connections, part II - hybrid connections, J. Math. Anal. Appl. 67(1979)215-231.

R. J. Duffin
Department of Mathematics
Carnegie Mellon University
Pittsburgh, Pennsylvania 15312

T. D. Morley
Department of Mathematics
University of Illinois
Urbana, Illinois 61901

This research was partially supported by Army Research Office, Research Triangle, N.C., Grant DA-AROD-31-124-73-G17. Some of the results appear in the second author's Ph.D. thesis at Carnegie-Mellon University.

C D MEYER Jr

Analysis of finite Markov chains by group inversion techniques

1. INTRODUCTION

Consider a general finite homogeneous Markov chain whose transition matrix T exhibits no special structure. The problem is to analyze the chain. A first step is to decide whether or not the chain is ergodic. If it is not, one then wishes to isolate and identify the distinct transient and ergodic classes. Having completed this, the next task is usually to determine the flow through the chain by reducing the transition matrix T to a block triangular form. Using this form, one proceeds to compute items such as probabilities for absorption into ergodic classes from transient states, mean absorption times into ergodic classes from transient states, etc.

The traditional approach is to first apply combinatorial techniques to obtain the permutation of states which will produce the block triangular form for T and then to numerically invert the relevant matrices obtained from the block triangular form in order to compute absorption times, absorption probabilities, etc.

The goal of this paper is to offer an alternate approach to this type of analysis.

For every chain, regardless of the structure, the matrix $A = I - T$ is singular but is always a member of a multiplicative matrix group [1], [2]. Thus A possesses an inverse, $A^{\#}$, with respect to this group [2]. It is our purpose to show that the entries of this group inverse matrix, $A^{\#}$, answer all relevant questions pretaining to the associated Markov chain. Furthermore, we will demonstrate that it is in fact <u>not necessary</u> to have first performed a combinatorial analysis in order to permute T into block triangular form. With a single computation, namely a group inversion, one extracts not only all of the important probabilistic aspects of the chain, but also (at no additional cost) all of the important combinatorial information concerning the connectivity of the chain as well. If one desires, the block triangular form for T can be produced as a simple by-product (completely free of charge) from the knowledge of $A^{\#}$.

2. INTERPRETING THE MEANING OF THE ENTRIES OF $A^{\#}$ AND $I - AA^{\#}$.

For the sake of clarity, the terminology used in the sequel is explicitly spelled out in the following definition.

Definition 2.1 For a finite homogeneous Markov chain with transition matrix $T = [p_{ik}](T^n = [p_{ik}^{(n)}])$:

(i) An irreducible class of states is a set of states in which every state of the set is accesible from every other state of the set. (i.e., for each pair of states S_i and S_k in the set, there is an integer $n > 0$ such that $p_{ik}^{(n)} > 0$.)

(ii) An ergodic class (or ergodic set) is an irreducible class of states such that once the process enters the class, it is impossible to leave. (An ergodic class is an irreducible subchain).

(iii) An ergodic chain is one whose states form a single ergodic class. (A chain is ergodic if and only if its transition matrix is an irreducible matrix).

(iv) A regular chain is a chain such that $T^n > 0$ for some positive integer n. (A regular chain is clearly ergodic).

(v) An ergodic state is a member of an ergodic class.

(vi) A transient class (or transient set) is an irreducible class of states with the property that it is possible for the process to leave the class.

(vii) A transient state is a member of a transient class.

This paper will deal only with finite homogeneous chains so that the phrase "finite homogeneous" will be understood, but not explicitly stated, when the term "chain" is used.

For a chain with transition matrix T, it is shown in [1] and [2] that $I\text{-}AA^{\#}$ always represents the limiting matrix of the chain in the sense that

$$I - AA^{\#} = \begin{cases} \lim_{n\to\infty} \dfrac{I + T + \ldots + T^{n-1}}{n} & \text{always holds} \\ \lim_{n\to\infty} T^n & \text{when the limit exists.} \end{cases} \quad (2.1)$$

The structure of this limiting matrix is important in analyzing a general Markov chain. Therefore, it is important to understand the structure of the matrices $A^{\#}$ and $I-AA^{\#}$.

Before analyzing the nature of the matrices $A^{\#}$ and $I-AA^{\#}$, the following facts are needed.

Lemma 2.1 If $S \geq 0$ is an irreducible substochastic matrix (i.e., each row sum is at most 1 with at least one row sum strictly less than 1), then $\rho(s) < 1$ so that $I-S$ is nonsingular and $(I-S)^{-1} > 0$. ($\rho(\cdot)$ denotes spectral radius).

This fact is well known and can be proven in several different ways. For the sake of completness, one proof is indicated below.

Proof Let $r = \rho(s)$. The Perron-Frobenius Theorem guarantees the existence of a (row) vector χ' such that $\chi' > 0$ and $\chi'S = r\chi'$ where $\sum \chi_i = 1$. The fact that S is substochastic means that

$$Sj = j - v \text{ for some } v \geq 0,\ v \neq 0, \text{ and } j = [1, \ldots, 1]'.$$

Multiplication on the left by χ' yields $r = 1 - \chi'v < 1$, which in turn implies that $I-S$ is nonsingular and $(I-S)^{-1} = \sum_{n=0}^{\infty} S^n$. The fact that $\sum_{n=0}^{\infty} S^n > 0$ is an immediate consequence of the irreducibility of S. □

Lemma 2.2 If A is a group matrix (i.e., $A^{\#}$ exists) and P is a permutation matrix such that

$$A = P' \begin{bmatrix} A_{11} & A_{12} \\ 0 & A_{22} \end{bmatrix} P$$

then A_{11} and A_{22} are group matrices such that $(I - A_{11}A_{11}^{\#})\ A_{12} \times (I - A_{22}A_{22}^{\#}) = 0$ and $A^{\#}$ and $I - AA^{\#}$ are given by

$$A^{\#} = P' \left[\begin{array}{c|c} A_{11}^{\#} & A_{11}^{\#2}A_{12}(I - A_{22}A_{22}^{\#}) + (I - A_{11}A_{11}^{\#})\, A_{12}A_{22}^{\#2} - A_{11}^{\#}A_{12}A_{22}^{\#} \\ \hline 0 & A_{22}^{\#} \end{array}\right] P \quad (2.2)$$

$$I - AA^{\#} = P' \left[\begin{array}{c|c} I-A_{11}A_{11}^{\#} & -A_{11}^{\#}A_{12}(I-A_{22}A_{22}^{\#}) - (I-A_{11}A_{11}^{\#})\, A_{12}A_{22}^{\#} \\ \hline 0 & I-A_{22}A_{22}^{\#} \end{array}\right] P \quad (2.3)$$

If A_{11} is nonsingular, then

$$A^{\#} = P' \left[\begin{array}{c|c} A_{11}^{-1} & A_{11}^{-2}A_{12}\,(I-A_{22}A_{22}^{\#}) - A_{11}^{-1}A_{12}A_{22}^{\#} \\ \hline 0 & A_{22}^{\#} \end{array}\right] P \quad (2.4)$$

and

$$I - AA^{\#} = P' \left[\begin{array}{c|c} 0 & -A_{11}^{-1}A_{12}\,(I-A_{22}A_{22}^{\#}) \\ \hline 0 & I-A_{22}A_{22}^{\#} \end{array}\right] P. \quad (2.5)$$

Proof One can easily verify the conditions of Definition 1.1 are satisfied. These results are also proven in more generality in [1] and [4]. □

Lemma 2.3 For every transition matrix T, there is a permutation matrix, P, such that

$$T = P' \left[\begin{array}{cccc|cccc} T_{11} & T_{12} & \cdots & T_{1r} & T_{1,r+1} & T_{1,r+2} & \cdots & T_{1n} \\ 0 & T_{22} & \cdots & T_{2r} & T_{2,r+1} & T_{2,r+2} & \cdots & T_{2n} \\ \vdots & & & & & & & \\ 0 & 0 & \cdots & T_{rr} & T_{r,r+1} & T_{r,r+2} & \cdots & T_{rn} \\ \hline 0 & 0 & \cdots & 0 & T_{r+1,r+1} & 0 & \cdots & 0 \\ 0 & 0 & \cdots & 0 & 0 & T_{r+2,r+2} & \cdots & 0 \\ \vdots & & & & & & & \\ 0 & 0 & \cdots & 0 & 0 & 0 & \cdots & T_{nn} \end{array}\right] P \qquad (2.6)$$

where each T_{ii}, $1 \leq i \leq r$, is an irreducible substochastic matrix and each T_{ii}, $i > r$, is an irreducible stochastic matrix. For $1 \leq i \leq r$, $\rho(T_{ii}) < 1$ and for $i < r$, T_{ii} is the transition matrix of an ergodic subchain.

Proof Let P be the permutation matrix corresponding to a permutation of the states so that all transient states are listed before any ergodic states and states belonging to common transient sets are listed sequentially and states belonging to common ergodic sets are listed sequentially. The fact that $\rho(T_{ii}) < 1$ for $1 \leq i \leq r$ follows from Lemma 2.1. □

3. INTERPRETING THE ENTRIES OF $A^{\#}$

Throughout this paper, the following notations will be used.

Definition 3.1 Let B be a $m \times m$ matrix and $\mathcal{C}$ be a chain with states $\{S_1, S_2, \ldots, S_m\}$. Let $\mathcal{S}_i = \{S_{i_1}, S_{i_2}, \ldots, S_{i_p}\}$ and $\mathcal{S}_k = \{S_{k_1}, S_{k_2}, \ldots, S_{k_q}\}$ be any two subsets of states.

(i) $B[i|k]$ denotes the (i,k)-entry of B.

(ii) $B[i|*]$ denotes the i-th row of B.

(iii) $B[*|k]$ denotes the k-th column of B.

(iv) $B[i|S_k]$ denotes those entries of B which lie on the inter-

(v) section of the i-th row with columns $k_1, k_2 \ldots, k_q$.

$B[S_i|S_k]$ denote those entries of B which lie on the inter-

section of rows $i_1, i_2, \ldots, i_p$ with columns $k_1, k_2, \ldots, k_q$.

It is easy to interpret the entries of $A^{\#}$ which correspond to the transient states. The first result along these lines is given below.

Theorem 3.1 If S_i and S_k are transient states, then

$A^{\#}[i|k]$ = the expected number of times the process is in S_k when initially in S_i.

Proof Use Lemma 2.3 to write T and A in the form

$$T = P' \left[\begin{array}{c|c} Q & R \\ \hline 0 & E \end{array}\right] P \quad \text{and} \quad A = P' \left[\begin{array}{c|c} I - Q & -R \\ \hline 0 & I - E \end{array}\right] P. \qquad (3.1)$$

Since $\rho(T_{ii}) < 1$ for $1 \leq i \leq r$, it follows that $\rho(Q) < 1$ so that $I - Q$ is nonsingular and

$$(I - Q)^{-1} = \sum_{p=0}^{\infty} Q^p.$$

If S_i and S_k are transient states, then the (i,k)-entry of T is equal to some entry (say the (r,s) - entry) of Q. Observe now that (2.4) says that

$$A^{\#}[i|k] = [\,(I-Q)^{-1}]\quad [r|s] \quad .$$

The expected number of times in S_k when initially in S_i is given by

$$\lim_{n\to\infty}\left[\sum_{p=0}^{n-1} T^p\right][i|k] = \lim_{n\to\infty}\left[\sum_{p=0}^{n-1} Q^p\right][r|s] = \left[(I-Q)^{-1}\right][r|s]$$

$$= A^{\#}[i|k]. \qquad \square$$

Perhaps one of the most important problems concerning a general chain which contains transient and ergodic states is that of determining the mean absorption time from a given initial state. The previous theorem makes it clear that this information is directly available from the entries of $A^{\#}$.

<u>Theorem 3.2</u> Let T denote the set of indicies corresponding to the transient states. If the process is initially in state S_i, then

$$\sum_{k\in T} A^{\#}[i|k] = \text{The mean number of steps to reach an ergodic set.}$$

<u>Proof</u> If S_i is a transient state, then the result is a direct consequence of Theorem 3.1. If S_i is an ergodic state, then write T as in (3.1) and observe from (2.4) that $\sum_{k\in T} A^{\#}[i|k] = 0$, which is the desired result in this case. $\square$

Finally, the following variances can be obtained from the entries of $A^{\#}$.

<u>Theorem 3.3</u> Let $A^{\#}_{dg}$ denote the diagonal matrix obtained from $A^{\#}$ by setting all off diagonal entries to zero and let $A^{\#}_{sq}$ denote the matrix obtained by squaring each entry of $A^{\#}$. If S_i and S_k are transient states then

$$[A^{\#}(2A^{\#}_{dg} - I) - A^{\#}_{sq}]\ [i|k] = \text{variance of the number of times in } S_k \text{ when initially in } S_i\,.$$

If T is the set of indicies corresponding to the transient states, then

$$\sum_{k\in T} [(2A^{\#} - I)A^{\#} - A^{\#}_{sq}]\ [i|k] = \text{variance of the number of steps until absorption when initially in } S_i.$$

The proof is straight forward by using the results of [2]. The details are left to the interested reader.

4. INTERPRETING THE ENTRIES OF $I-AA^{\#}$

The importance of the matrix $I-AA^{\#}$ lies in the fact that it is the limiting matrix in the sense of (2.1). As such, the individual entries represent limiting probabilities, but possibly in a **generalized** sense.

The purpose of this section is to develop the explicit interpretation of each entry of $I-AA^{\#}$ by utilizing only a few elementary properties of the group inverse. First, some notation and terminology is needed.

Definition 4.1 For any state S_k of a chain, $[S_k]$ will denote the irreducible class of states which contains S_k. If S_k is an ergodic state and $[S_k] = \{S_\alpha, S_\beta, \ldots, S_k, \ldots, S_j\}$ with $\alpha < \beta < \ldots < k < \ldots < j$, then there is a limiting probability vector $w' = (w_\alpha, w_\beta, \ldots, w_k \ldots, w_j)$ associated with the ergodic subchain $[S_k]$. The component w_k is called the *limiting probability associated with the state* S_k.

The fundamental theorem regarding the individual entries of $I - AA^{\#}$ can be stated as follows:

Theorem 4.1 The (i, k) - entry of $I-AA^{\#}$ is given by

$$(I-AA^{\#})[i|k] = \begin{cases} 0 \text{ when } S_k \text{ is a transient state.} \\ w_k \cdot P(\text{absorption into } [S_k] \text{ initially in } S_i) \text{ when } S_k \\ \text{is an ergodic state where } w_k \text{ is the limiting} \\ \text{probability associated with } S_k. \end{cases}$$

Proof Write the transition matrix as

$$T = P' \begin{bmatrix} T_{11} & T_{12} & T_{13} \\ 0 & T_{22} & 0 \\ 0 & 0 & T_{33} \end{bmatrix} P \quad P \text{ a permutation matrix} \qquad (4.1)$$

where T_{22} is the transition matrix corresponding to $[S_k]$ and T_{33} is the transition matrix corresponding to all other ergodic states. T_{11} corresponds to all transient states. Using Lemma 2.2, it is easy to deduce that

$$I-AA^{\#} = P'\left[\begin{array}{c|c|c} 0 & -A_{11}^{-1}A_{12}\,(I-A_{22}A_{22}^{\#}) & -A_{11}^{-1}A_{13}\,(I-A_{33}A_{33}^{\#}) \\ \hline 0 & I-A_{22}A_{22}^{\#} & 0 \\ \hline 0 & 0 & I-A_{33}A_{33}^{\#} \end{array}\right] P \tag{4.2}$$

Proceed by considering the various possible cases. Suppose first that S_k is a transient state.

S_k transient: It is clear that if S_k is a transient state, then $(I-AA^{\#})$-$[i|k]$ lies in either the (1,1)-block, the (2,1)-block or the (3,1)-block of the permuted matrix $P(I-AA^{\#})P'$ and hence $(I-AA^{\#})\,[i|k] = 0$.

Suppose now that S_k is an ergodic state. State S_i can be either ergodic or transient. First consider the case where S_i is ergodic.

S_i and S_k ergodic: If $S_i \in [S_k]$ then P(absorption into $[S_k]$ | initially in S_i) = 1. From (4.2) it is clear that $(I-AA^{\#})[i|k] = (I-A_{22}A_{22}^{\#})[r|s]$ for some r and s. By (2.1), $I-A_{22}A_{22}^{\#}$ is the limiting matrix for $[S_k]$ so that every element in the s-th column of $I-A_{22}A_{22}^{\#}$ is equal to w_k, and the desired result follows. If $S_i \notin [S_k]$ then P(absorption into $[S_k]$ | initially in S_i) = 0 and $(I-AA^{\#})\,[i|k]$ lies in the (3,2)-block of $P(I-AA^{\#})P'$ so that the desired result is again obtained.

Now consider the case where S_i is a transient state.

S_i transient and S_k ergodic: In this case, $(I-AA^{\#})\,[i|k]$ lies in the (1,2)-block of $P(I-AA^{\#})P'$. Say that $(I-AA^{\#})\,[i|k] =$ $[-A_{11}^{-1}A_{12}(I-A_{22}A_{22}^{\#})]\,[r|s]$ and let e_r, e_s, and j denote the r-th unit vector, the s-th unit vector, and the vector of all 1's, respectively. Using the fact that $I-A_{22}A_{22}^{\#}$ is the limiting matrix for $[S_k]$, it can be seen that

$$(I-AA^{\#})[i|k] = -e_r'\,A_{11}^{-1}A_{12}(I-A_{22}A_{22}^{\#})\,e_s = -e_r'A_{11}^{-1}A_{12}(w_k j) = w_k e_r'(I-T_{11})^{-1}T_{12}j.$$

By converting each ergodic state into an absorbing state, it is easily deduced from the well known theory of absorbing chains that

$e_r'(I-T_{11})^{-1}T_{12}j$ = P(absorption into $[S_k]$| initially in S_i), so that the theorem is proven. □ (4.3)

For a general chain with more than one ergodic set, it is usually necessary to determine the probability of eventual absorption into a particular ergodic set from a given initial state. The previous theorem makes it clear that this information is directly available from the entries of $I-AA^{\#}$.

Theorem 4.2 Let $E = S_{k_1}, S_{k_2}, \ldots, S_{k_m}$ be an ergodic set of states and let $I = k_1, k_2, \ldots, k_m$ be the set of indicies of those states which belong to E. For any initial state,

$$\sum_{k\in I} (I-AA^{\#})[i|k] = P(\text{absorption into } E \mid \text{initially in } S_i).$$

Proof The proof follows from Theorem 4.1 by observing that

$\sum_{k\in I} w_k = 1.$ □

An obvious corollary to Theorems 4.1 and 4.2 is the application to the situation where each ergodic set is a singleton. (i.e., application to absorbing chains).

Corollary 4.1 If S_k is an absorbing state of an absorbing Markov chain, then

$$(I-AA^{\#})[i|k] = P(\text{absorption into } S_k \mid \text{initially in } S_i).$$

For a general chain whose n-step transition probabilities are $p_{ik}^{(n)}$, the limit $\lim_{n\to\infty} p_{ik}^{(n)}$ may not exist for all values of i and k. However, this limit may exist for some values of i and k and when it exists, it is an entry of $I-AA^{\#}$. In fact, it is easy to prove that

$$\lim_{n \to \infty} p_{ik}^{(n)} \text{ exists if and only if } \begin{cases} S_k \text{ is a transient state} \\ \qquad \text{or} \\ S_k \text{ is an ergodic state and } [S_k] \text{ is} \\ \text{a regular subchain.} \end{cases} \tag{4.4}$$

Now, (2.1) guarantees that it is always the case that

$$(I-AA^{\#})\,[i|k] = \lim_{n \to \infty} \frac{\delta_{ik} + p_{ik} + p_{ik}^{(2)} + \ldots + p_{ik}^{(n-1)}}{n}, \tag{4.5}$$

$$\delta_{ik} = \begin{cases} 0 & i \neq k \\ 1 & i = k \end{cases}.$$

When the limit (4.4) exists, it must agree with the Cesaro limit (4.5) so that one has a simple proof of the following corollary.

Corollary 4.2 If $\lim_{n \to \infty} p_{ik}^{(n)}$ exists, then it is given by

$$\lim_{n \to \infty} p_{ik}^{(n)} = (I-AA^{\#})[i|k] = \begin{cases} 0 \text{ when } S_k \text{ is a transient state} \\ \\ w_k \, P(\text{absorption into } [S_k] \,|\, \text{initially in } S_i) \\ \text{when } S_k \text{ is an ergodic state} \end{cases}$$

where w_k is the limiting probability associated with S_k in the regular subchain $[S_k]$.

5. USING $A^{\#}$ AND $I-AA^{\#}$ TO ANALYZE THE STRUCTURE OF THE CHAIN AND TO CLASSIFY THE STATES

In order to use the results of the traditional theory to analyze chains of the type under consideration here, one must first be able to distinguish the transient states from the ergodic states and be able to recognize the disjoint irreducible classes before any calculations can be made. That is, one essentially needs to know the canonical form of the transition matrix as given in Lemma 2.3.

The problem of classification is usually solved by graph theoretic techniques. Once the canonical form of the transition matrix is apparent, the necessary matrix inversions can then be performed to obtain the information regarding the absorption probabilities, expected values, and variances which were discussed in the previous section.

The beauty of the group inverse as it is used in the present context, lies in the fact that it is not necessary to first perform the analysis of state classification. By a single (group) matrix inversion, (i.e., by obtaining $A^{\#}$) one can immediately distinguish the transient states from the ergodic states as well as isolate the irreducible classes and detect the flow of the process by simply inspecting the algebraic signs of the entries in the matrices $A^{\#}$ and $I-AA^{\#}$. At the same time, one also has obtained the limiting properties along with the other relevant information that one usually wishes to know.

This section is devoted to showing how the problem of state classification is automatically solved once $A^{\#}$ is known.

The first result shows how the matrix $I-AA^{\#}$ can be used to determine whether or not a chain is an ergodic chain, and hence whether or not the transition matrix is an irreducible matrix.

Theorem 5.1 A Markov chain is an ergodic chain if and only if $I-AA^{\#} > 0$.

Proof A chain is ergodic if and only if the transition matrix, T, is irreducible, which in turn is equivalent to saying that

$$\lim_{n \to \infty} \frac{I + T + T^2 + \ldots + T^{n-1}}{n} > 0 .$$

The desired conclusion is now immediate from 2.1. □

The problem of distinguishing the transient states from the ergodic states can be solved by inspecting either $A^{\#}$ or $I-AA^{\#}$. Below is the result which tells how to use $A^{\#}$. One simply notes the positions of the nonnegative columns in $A^{\#}$.

Theorem 5.2 State S_k is a transient state if and only if

$A^{\#}[*|k] \geq 0$ and $A^{\#}[i|k] > 0$ for some i.

<u>Proof</u> Write T and A as

$$T = P' \left[\begin{array}{c|c} T_{11} & T_{12} \\ \hline 0 & T_{22} \end{array}\right] P, \qquad A = P' \left[\begin{array}{c|c} A_{11} & A_{12} \\ \hline 0 & A_{22} \end{array}\right] P$$

where T_{11} corresponds to the transient states and T_{22} corresponds to the ergodic states so that $A^{\#}$ is given by (2.4). Assume first that S_k is a transient state. This means that the elements of $A^{\#}[*|k]$ can all be found in some column of the matrix $\left[\begin{array}{c} A_{11}^{-1} \\ \hline 0 \end{array}\right]$ but in a permuted order.

It follows from Lemmas 2.1 and 2.3 that $A_{11}^{-1} = (I-T_{11})^{-1} \geq 0$. Since A_{11}^{-1} has no zero columns, it is clear that $A^{\#}[*|k] \geq 0$ with at least one positive entry. Conversely, suppose that $A^{\#}[*|k] \geq 0$ with at least one positive entry. Write T as

$$T = P' \left[\begin{array}{c|cccc} T_{11} & T_{12} & T_{13} & \cdot\cdot & T_{1n} \\ \hline 0 & T_{22} & 0 & \cdot\cdot & 0 \\ 0 & 0 & T_{33} & \cdot\cdot & 0 \\ \cdot & & & \cdot\cdot & \\ 0 & 0 & 0 & \cdot\cdot & T_{nn} \end{array}\right] P, \text{ P a permutation,} \tag{5.1}$$

where each T_{ii}, $i \geq 2$, is the transition matrix of an ergodic class and T_{11} corresponds to the transient states. It follows from (2.4) that

$$A^{\#} = P' \left[\begin{array}{c|c|c|c} A_{11}^{-1} & A_{11}^{-2}A_{12}(I-A_{22}A_{22}^{\#})-A_{11}^{-1}A_{12}A_{22}^{\#} & \cdot\cdot & A_{11}^{-2}A_{1n}(I-A_{nn}A_{nn}^{\#})-A_{11}^{-1}A_{1n}A_{nn}^{\#} \\ \hline 0 & A_{22}^{\#} & \cdot\cdot & 0 \\ \hline \cdot & \cdot & \cdot\cdot & \cdot \\ \hline 0 & 0 & \cdot\cdot & A_{nn}^{\#} \end{array}\right] P$$

The object is to prove S_k is a transient state. Suppose this is not the case. (i.e., assume S_k is an ergodic state). This means that the entries of $A^{\#}[*|k]$ all lie in some column of a matrix of the form

$$B = \begin{bmatrix} A_{11}^{-2}A_{1r}(I-A_{rr}A_{rr}^{\#})-A_{11}^{-1}A_{1r}A_{rr}^{\#} \\ 0 \\ \cdot \\ 0 \\ A_{rr}^{\#} \\ 0 \\ \cdot \\ 0 \end{bmatrix}$$

but in permuted order. Say the entries of $A^{\#}[*|k]$ are all found in the i-th column, $B[*|i]$, of B. Write $B[*|i]$ as

$$B[*|i] = \begin{bmatrix} A_{11}^{-2}A_{1r}(I-A_{rr}A_{rr}^{\#})[*|i] - A_{11}^{-1}A_{1r}A_{rr}^{\#}[*|i] \\ 0 \\ \cdot \\ 0 \\ A_{rr}^{\#}[*|i] \\ 0 \\ \cdot \\ 0 \end{bmatrix} \geq 0 \text{ with at least one positive entry.} \qquad (5.2)$$

Claim: It is claimed that $A_{rr}^{\#}[*|i] \neq 0$.

Proof If $A_{rr}^{\#}[*|i] = 0$, then

$$B[*|i] = \begin{bmatrix} A_{11}^{-2}A_{1r}[*|i] \\ \\ 0 \\ \cdot \\ 0 \end{bmatrix}.$$

Since $A_{11}^{-1} \geq 0$ and $A_{1r} \leq 0$, it is impossible for $B[*|i]$ to have a positive entry. However, the hypothesis guarantees that $B[*|i]$ always has at least one positive entry. Therefore, $A_{rr}^{\#}[*|i]$ must have a nonzero component so that the claim is proven. It is now clear from (5.2) that $A_{rr}^{\#}[*|i] \geq 0$ with at least one positive entry. It is also clear from Theorem 5.1 that $I-A_{rr}A_{rr}^{\#} > 0$. One can now conclude that $(I-A_{rr}A_{rr}^{\#})A_{rr}^{\#}[*|i] > 0$. However, this is impossible because $(I-A_{rr}A_{rr}^{\#})A_{rr}^{\#} = 0$. Therefore the supposition that S_k is an ergodic state is false so that S_k must be a transient state. □

It is also extremely easy to inspect the matrix $I-AA^{\#}$ and immediately distinguish the transient states from the ergodic states.

Theorem 5.3 State S_k is a transient state if and only if $(I-AA^{\#})[*|k] = 0$. (Equivalently, S_k is an ergodic state if and only if the k-th column has at least one nonzero component).

Proof Write T as in (5.1). From (2.5) it can be seen that $I-AA^{\#}$ has the form

$$I-AA^{\#} = \left[\begin{array}{c|ccccc} 0 & * & * & \cdot & * \\ \hline 0 & I-A_{22}A_{22} & 0 & \cdot & 0 \\ 0 & 0 & I-A_{33}A_{33} & \cdot & 0 \\ \cdot & & & \cdot & \cdot \\ \cdot & & & \cdot & \cdot \\ 0 & 0 & 0 & \cdot & I-A_{nn}A_{nn}^{\#} \end{array}\right]. \tag{5.3}$$

For $i \geq 2$, T_{ii} is the transition matrix for an ergodic chain so that Theorem 5.1 guarantees that $I-A_{ii}A_{ii}^{\#} > 0$. It is now clear from (5.3) that state S_k is transient if and only if the k-th column of $I-AA^{\#}$ is entirely zero. □

Now that the problem of distinguishing between the transient states and the ergodic states has been solved, it is necessary to isolate the distinct irreducible sets. That is, identify the transient classes as well as the ergodic classes and then order them.

It will first be demonstrated how to use $A^{\#}$ to determine the transient classes. Following this, a discussion of how to determine the ergodic classes will be given.

Theorem 5.4 Let S_i and S_k be transient states. These states belong to the same transient set if and only if both $A^{\#}[i|k] > 0$ and $A^{\#}[k|i] > 0$.

Proof Suppose first that S_i and S_k belong to the same transient set. This means that the expected number of times the process is in S_k when initially in S_i, as well as the expected number of times the process is in S_i when initially in S_k, is positive. The fact that $A^{\#}[i|k] > 0$ and $A^{\#}[k|i] > 0$ now follows from Theorem 3.1 Conversely, suppose that $A^{\#}[i|k]$ and $A^{\#}[k|i]$ are both positive. Theorem 3.1 now guarantees that the process can move from S_i into S_k as well as being able to move from S_k into S_i. Therefore, S_i and S_k must belong to the same irreducible class, and hence the same transient set since both states are transient. □

The next result is an immediate corollary to the previous Theorem.

Corollary 5.1 Let T a collection of transient states. The states of T each belong to the same transient class if and only if $A^{\#}[T|T] > 0$.

Now address the problem of counting and isolating the distinct ergodic classes as well as the distinct transient classes. For the sake of convenience, the following notation and terminology is used in the sequel.

Definition 5.1 The ergodic index of a finite chain $\mathcal{C}$ is defined to be the number of distinct ergodic classes contained in $\mathcal{C}$. The ergodic index of $\mathcal{C}$ is denoted by $\mathcal{E}(\mathcal{C})$. The transient index of $\mathcal{C}$ is defined to be the number of distinct transient classes contained in $\mathcal{C}$ and is denoted by $\mathcal{T}(\mathcal{C})$.

The problem of determing the ergodic index is easily solved once $I-AA^{\#}$ or $A^{\#}$ is known.

Theorem 5.5 If $\mathcal{C}$ is a chain with m states, then $\mathcal{E}(\mathcal{C})$ is given by each of the following expressions.

$$\begin{aligned}\mathcal{E}(\mathcal{C}) &= \text{Trace } (I-AA^{\#}) &&= \text{Rank } (I-AA^{\#})\\ &= m - \text{Trace } (AA^{\#}) &&= m - \text{Rank } (AA^{\#})\\ &= m - \text{Rank } (A^{\#}) &&= m - \text{Rank } (A)\\ &= \dim N(A) &&= \text{algebraic multiplicity of 1 as an eigenvalue of } T.\end{aligned}$$

Proof From (2.6) it is seen that if $\mathcal{E}(\mathcal{C}) = r$, then T has the form

$$T = P' \left[\begin{array}{c|cccc} T_{00} & T_{01} & T_{02} & \cdot\cdot & T_{0r} \\ \hline 0 & T_{11} & 0 & \cdot\cdot & 0 \\ 0 & 0 & T_{22} & \cdot\cdot & 0 \\ \cdot & & & & \\ 0 & 0 & 0 & \cdot\cdot & T_{rr} \end{array}\right] P$$

where each T_{ii}, $i \geq 1$, corresponds to one of the ergodic classes. From (2.5) it follows that $I-AA^{\#}$ has the form

$$I-AA^{\#} = P' \left[\begin{array}{c|cccc} 0 & * & * & \cdot\cdot & * \\ \hline 0 & I-A_{11}A_{11}^{\#} & 0 & \cdot\cdot & 0 \\ 0 & 0 & I-A_{22}A_{22}^{\#} & \cdot\cdot & \\ \cdot & & & & \\ \cdot & & & & \\ 0 & 0 & 0 & \cdot\cdot & I-A_{rr}A_{rr}^{\#} \end{array}\right] P$$

It is guaranteed by (2.1) that $I-A_{ii}A_{ii}^{\#}$ represents the limiting matrix (in permuted form) for the i-th ergodic class, $i \geq 1$. Furthermore, $I-A_{ii}A_{ii}^{\#}$ is idempotent so that Trace $(I-A_{ii}A_{ii}^{\#})$ = Rank $(I-A_{ii}A_{ii}^{\#}) = 1$. Therefore,

$$\text{Rank } (I-AA^{\#}) = \text{Trace } (I-AA^{\#}) = \sum_{i=1}^{r} \text{Trace } (I-A_{ii}A_{ii}^{\#}) = r.$$

This establishes the first two equalities. The remaining equalities are straightforward. ☐

Not only does $I-AA^{\#}$ provide an immediate answer as to how large $E(C)$ is, but $I-AA^{\#}$ also distinguishes the distinct ergodic classes from one another.

Theorem 5.6 If S_i and S_k are both ergodic states, then the following statements are equivalent.

States S_i and S_k belong to the same ergodic class. (5.4)

$(I-AA^{\#})\ [i|k] > 0$ (Recall from (2.1) that $I-AA^{\#} \geq 0$) (5.5)

$(I-AA^{\#})\ [i|*] = (I-AA^{\#})\ [k|*]$ (5.6)

Proof Write the transition matrix as in (5.1) so that $I-AA^{\#}$ has the form of that in (5.3). For $r \geq 2$, T_{rr} is the transition matrix of an ergodic chain. The limiting matrix for any ergodic chain has identical rows and strictly positive entries. Equation (2.1) guarantees that $I-A_{rr}A_{rr}^{\#}$ is the limiting matrix for T_{rr} so that the proof of the theorem is now immediate. ☐

Although it is extremely easy to use the matrix $I-AA^{\#}$ to isolate the distinct ergodic classes, it should be observed that the matrix $A^{\#}$, by itself, can also aid in this purpose. This is brought to light by the following result.

Theorem 5.7 Let S_i and S_k be two ergodic states. If either $A^{\#}[i|k] \neq 0$ or $A^{\#}[k|i] \neq 0$, then S_i and S_k belong to the same ergodic class.

Proof Prove the contrapositive is true. That is, show that if S_i and S_k belong to different ergodic classes, then $A^{\#}[i|k] = 0$ and $A^{\#}[k|i] = 0$. Without loss of generality, one can assume the states have been arranged so that the transition matrix has the form

$$T = \left[\begin{array}{c|cccc} T_{00} & T_{01} & T_{02} & \cdot\cdot & T_{0r} \\ \hline 0 & T_{11} & 0 & \cdot\cdot & 0 \\ 0 & 0 & T_{22} & \cdot\cdot & 0 \\ \cdot & & & \cdot\cdot & \cdot \\ 0 & 0 & 0 & \cdot\cdot & T_{rr} \end{array}\right]$$

where each T_{ii}, $i \geq 1$ corresponds to a distinct ergodic class. The matrix $A^{\#}$ then has the form

$$A^{\#} = \left[\begin{array}{c|cccc} A_{00}^{-1} & * & * & \cdot\cdot & * \\ \hline 0 & A_{11}^{\#} & 0 & \cdot\cdot & 0 \\ 0 & 0 & A_{22}^{\#} & \cdot\cdot & 0 \\ \cdot & & & & \\ \cdot & & & & \\ 0 & 0 & 0 & \cdot\cdot & A_{rr}^{\#} \end{array}\right].$$

From this form, it is clear that if S_i and S_k belong to different ergodic classes, then $A^{\#}[i|k] = 0$ and $A^{\#}[k|i] = 0$. □

Unfortunately, the converse of this Theorem is false. The chain whose transition matrix is

$$T = \frac{1}{8}\begin{bmatrix} 3 & 1 & 4 \\ 1 & 6 & 1 \\ 4 & 1 & 3 \end{bmatrix}$$

is an ergodic chain so that states S_1 and S_3 definitely belong to the same ergodic class, However,

$$A^{\#} = \frac{8}{9}\begin{bmatrix} 1 & -1 & 0 \\ -1 & 2 & -1 \\ 0 & -1 & 1 \end{bmatrix}$$

so that $A^{\#}[1|3] = A^{\#}[3|1] = 0$

By simply observing the sign patterns of $A^{\#}$ one can easily count the number of distinct transient classes. For an arbitrary matrix M, let sgn(M) denote the <u>sign matrix</u> associated with M. That is,

$$[\mathrm{sgn}(M)]_{ii} = \begin{cases} +1 & \text{if } m_{ij} > 0 \\ 0 & \text{if } m_{ij} = 0 \\ -1 & \text{if } m_{ij} < 0 \end{cases}$$

<u>Theorem 5.8</u> If C is a finite Markov chain, then $T(C)$ is simply the number of distinct nonzero nonnegative columns of $\mathrm{sgn}(A^{\#})$.

<u>Proof</u> Suppose there are exactly k distinct transient classes. Then, from (2.6), T must have the form

$$T = P' \left[\begin{array}{cccc|c} T_{11} & T_{12} & \cdot\cdot & T_{1k} & * \\ 0 & T_{22} & \cdot\cdot & T_{2k} & * \\ \cdot & & & & \\ \cdot & & & & \\ 0 & 0 & \cdot\cdot & T_{kk} & * \\ \hline 0 & 0 & \cdot\cdot & 0 & E \end{array}\right] P$$

where each T_{ii} is irreducible and substochastic. By Lemma 2.3, $\rho(T_{ii}) < 1$ so that $A_{ii} = (I - T_{ii})$ is nonsingular. Moreover, the expansion

$A_{ii}^{-1} = \sum_{n=0}^{\infty} T_{ii}^{n}$ together with the fact that $T_{ii} \geq 0$ and T_{ii} is irreducible leads to the conclusion that $A_{ii}^{-1} > 0$ (i.e., A_{ii}^{-1} has only strictly positive entries).

Let

$$B_k = \begin{bmatrix} \beta_{11} & \beta_{12} & \cdot\cdot & \beta_{1k} \\ 0 & \beta_{22} & \cdot\cdot & \beta_{2k} \\ \cdot & & & \\ \cdot & & & \\ 0 & 0 & \cdot\cdot & \beta_{kk} \end{bmatrix} = \begin{bmatrix} A_{11} & A_{12} & \cdot\cdot & A_{1k} \\ 0 & A_{22} & \cdot\cdot & A_{2k} \\ \cdot & & & \\ \cdot & & & \\ 0 & 0 & \cdot\cdot & A_{kk} \end{bmatrix}^{-1}$$

where $A_{ii} = I - T_{ii}$ and $A_{ij} = -T_{ij}$ for $i \neq j$. It has already been established that $\beta_{ii} > 0$. The off diagonal blocks of B_k need not be strictly positive. However, if an off diagonal block β_{ij} contains at least one zero entry, then the entire block must be a zero matrix. This claim is substantiated below.

<u>Claim</u> For every i and j, either $\beta_{ij} > 0$ or else $\beta_{ij} = 0$.

<u>Proof of Claim:</u> Induct on k. For k = 1, there is nothing to prove.

For k = 2: $$B_2 = \begin{bmatrix} A_{11}^{-1} & -A_{11}^{-1}A_{12}A_{22}^{-1} \\ & \\ 0 & A_{22}^{-1} \end{bmatrix}$$

If $A_{12} \neq 0$, then $\beta_{12} - A_{11}^{-1}A_{12}A_{22}^{-1} > 0$ because $A_{11}^{-1} > 0$ and $A_{22}^{-1} > 0$. If $A_{12} = 0$, then $\beta_{12} = 0$.

Now assume the result is true for all matrices with k = n irreducible diagonal blocks and prove the result must hold for k = n + 1.

Write $$B_{n+1} = \left[\begin{array}{ccc|c} A_{11} & \cdot\cdot & A_{1n} & A_{1,n+1} \\ \cdot & & & \\ \cdot & & & \\ 0 & \cdot\cdot & A_{nn} & A_{n,n+1} \\ \hline 0 & \cdot\cdot & 0 & A_{n+1,n+1} \end{array}\right]^{-1} = \left[\begin{array}{c|c} B_n & -B_n \begin{bmatrix} A_{1,n+1} \\ \cdot \\ \cdot \\ A_{n,n+1} \end{bmatrix} \beta_{n+1,n+1} \\ \hline 0 & \beta_{n+1,n+1} \end{array}\right]$$

The induction hypothesis guarantees that $\beta_{ik} > 0$ or else $\beta_{ik} = 0$ for every

$i = 1, 2, \ldots, n + 1$ and $k = 1, 2, \ldots, n$. Consider $\beta_{i,\, n+1}$ for $i < n+1$.

$$\beta_{i,n+1} = -\sum_{\ell\, i}^{n} \beta_{i\ell} A_{\ell,n+1} \beta_{n+1,n+1}$$

Since each $\beta_{i\ell}$ as well as $\beta_{n+1,n+1}$ is either strictly positive or else entirely zero, it is clear that each term $-\beta_{i\ell} A_{\ell,n+1} \beta_{n+1}$ will be either strictly positive or else entirely zero so that $\beta_{i,n+1} > 0$ or else $\beta_{i,n+1} = 0$ for all i. Therefore, the result holds for $k = n$ and the claim is proven. From (2.4), $A^{\#}$ has the form

$$A^{\#} = P' \left[\begin{array}{c|c} B_k & C \\ \hline 0 & (I-E)^{\#} \end{array}\right] P .$$

By the nature of B_k (i.e., each $\beta_{ij} > 0$ or else $\beta_{ij} = 0$) it is clear that there are exactly k distinct nonzero nonnegative vectors contained in $\text{sgn}(B_k)$; one for each transient class. Therefore, there are at least k distinct nonzero nonnegative vectors contained in $\text{sgn}(A^{\#})$. All that remains to be proven is that there are no more than k such vectors. If there did exist more than k distinct nonzero nonnegative vectors in $\text{sgn}(A^{\#})$, then some column of $\left[\begin{array}{c} C \\ \hline (I-E)^{\#} \end{array}\right]$ would have to be nonnegative and nonzero. However, Theorem 5.2 would imply that such a column would correspond to a transient state, which is impossible because the transient states correspond to only columns in $\left[\begin{array}{c} B_k \\ \hline 0 \end{array}\right]$. Therefore, every column of $\left[\begin{array}{c} C \\ \hline (I-E)^{\#} \end{array}\right]$ must either have a negative entry or else be entirely zero and hence there are exactly k distinct nonzero nonnegative columns in $\text{sgn}(A^{\#})$. □

The above argument not only shows how to obtain $T(C)$ from $\text{sgn}(A^{\#})$, but it also demonstrates that $\text{sgn}(A^{\#})$ can be used to isolate the distinct transient classes.

Theorem 5.9 Let S_i and S_k be two transient states. S_i and S_k belong to the same transient class if and only if $\mathrm{sgn}(A^{\#}[*|i]) = \mathrm{sgn}(A^{\#}[*|k])$.

The proof follows directly from the proof of Theorem 5.8.

6. DETERMINING THE FLOW OF THE CHAIN

The problem of distinguishing between the transient states and ergodic states as well as isolating the distinct transient sets and ergodic sets is now completely solved once $A^{\#}$ is known. As far as state classification is concerned, the only remaining aspect is the analysis of the eventual flow of the process through the transient classes into the ergodic classes.

Let $\mathcal{S}_i = \{S_{i_1}, S_{i_2}, \ldots, S_{i_p}\}$ and $\mathcal{S}_k = \{S_{k_1}, S_{k_2}, \ldots S_{k_q}\}$ be two irreducible sets of states. It is clear that if one wishes to determine whether or not it is possible to move from $\mathcal{S}_i$ into $\mathcal{S}_k$ in a single step, he simply inspects the transition matrix. Movement from $\mathcal{S}_i$ into $\mathcal{S}_k$ in a single step is possible if and only if $T[\mathcal{S}_i|\mathcal{S}_k] \neq 0$.

However, if one wishes to decide on the possibility of eventual movement from $\mathcal{S}_i$ into $\mathcal{S}_k$, then many powers of T must be considered or else some other combinitorical technique is employed.

The purpose of this section is to show that the problem of determing the eventual flow of the chain is completely solved once $A^{\#}$ is known.

The first result along these lines presents conditions which insure the possibility of eventual movement from one state to another.

Definition 6.1 Let S_i and S_k be states of a chain whose transition matrix is T. State S_k is said to be accessible from S_i if $T^n[i|k] > 0$ for some n.

Theorem 6.1 Let S_i and S_k be any two states of a Markov chain. If $A^{\#}[i|k] \neq 0$, then S_k is accessible from S_i.

Proof It is well known that $A^{\#}$ can always be written as

$$A^{\#} = c_m A^m + c_{m-1} A^{m-1} + \ldots + c_1 A + c_o I$$

where m is less than the size of A. (See [1] or [2]). Therefore, if $A^{\#}[i|k] \neq 0$, then $A^{n}[i|k] \neq 0$ for some n. This means that there is at least one sequence of terms such that

$$A[i|r_1]\, A[r_1|r_2 \dots A[r_{n-2}|r_{n-1}]\, A[r_{n-1}|k] \neq 0$$

so that $A[i|r_1] \neq 0$, $A[r_1|r_2] \neq 0, \dots, A[r_{n-1}|k] \neq 0$. This in turn implies that in the directed graph, G(A) associated with A, there is a sequence of paths leading from node i to node k.

However, with the exception of loops, G(T) is the same as G(A) so that there is a sequence of paths from node i to node k in G(T) and hence the process can eventually move from state S_i to state S_k with a positive probability. □

Unfortunately, the converse of this theorem is not true. Consider a chain with transition matrix

$$T = \begin{bmatrix} 0 & 0 & 1 \\ 0 & \frac{1}{2} & \frac{1}{2} \\ 0 & \frac{1}{2} & \frac{1}{2} \end{bmatrix} \quad \text{so that} \quad A^{\#} = \begin{bmatrix} 1 & 1 & 0 \\ 0 & \frac{1}{2} & -\frac{1}{2} \\ 0 & -\frac{1}{2} & \frac{1}{2} \end{bmatrix}$$

The process can move from state S_1 to state S_3 but $A^{\#}[1|3] = 0$.

The fact that the converse of the previous theorem is false means that it is less than satisfying as far as determining the eventual flow of the process is concerned. This is remedied by the next theorem which can be used to provide a complete solution. First, a small lemma and then some terminology is needed.

Lemma 6.1 For a chain whose transition matrix is T, $A^{\#}[k|k] > 0$ if and only if $T[k|k] \neq 1$.

Proof Assume first that $T[k|k] \neq 1$. Let T be written in the form (2.6). From Lemma 2.2 it can be seen that $A^{\#}$ has the form

$$A^{\#} = P' \left[\begin{array}{ccccc|cccc} A_{11}^{-1} & * & * & . & * & * & * & & * \\ 0 & A_{22}^{-1} & * & . & * & * & * & & * \\ . & & . & & & & & & \\ 0 & 0 & 0 & . & A_{rr}^{-1} & * & * & & * \\ \hline 0 & 0 & 0 & . & 0 & A_{r+1,r+1}^{\#} & 0 & . & 0 \\ 0 & 0 & 0 & . & 0 & 0 & A_{r+2,r+2}^{\#} & . & 0 \\ 0 & 0 & 0 & . & 0 & 0 & 0 & . & A_{nn}^{\#} \end{array}\right] P.$$

Lemma 2.1 guarantees that $A_{ii}^{-1} > 0$ for $1 \leq i \leq r$. For $i > r$, T_{ii} must be the transition matrix of an ergodic chain with at least two states so that Theorem 4.1 of [2] establishes the fact that the diagonal elements of $A_{ii}^{\#}$ are positive for $i > r$. Conversely, if $A^{\#}[k|k] > 0$, then $T[k|k] \neq 1$. Otherwise, S_k is an ergodic class in itself and $A[k|k] = 0$. Clearly this would force $A^{\#}[k|k] = 0$. □

<u>Definition 6.2</u> Let S_i be same state of a chain with transition matrix T and let $\mathcal{S}_k$ be some irreducible class of states. The class $\mathcal{S}_k$ is said to be accessible from state S_i if $T^n[i|\mathcal{S}_k] \neq 0$ for some n.

The following result can now be established.

<u>Theorem 6.2</u> Let S_i be any state of a chain and let $\mathcal{S}_k$ be any irreducible class such that $\mathcal{S}_k \neq \{S_i\}$. The class $\mathcal{S}_k$ is accessible from state S_i if and only if $A^{\#}[i|\mathcal{S}_k] \neq 0$.

<u>Proof</u> Let $\mathcal{S}_k = \{S_{k_1}, S_{k_2}, \ldots, S_{k_n}\}$ and assume first that $A^{\#}[i|\mathcal{S}_k] \neq 0$. This means $A^{\#}[i|k_r] \neq 0$ for some r, $1 \leq r \leq n$. Theorem 6.1 now guarantees that the process can eventually move from state S_i to state $S_{k_r} \in \mathcal{S}_k$ with a positive probability. Conversely, suppose eventual movement from S_i into $\mathcal{S}_k$ is possible. The proof that $A^{\#}[i|\mathcal{S}_k] \neq 0$ is broken into three cases.

Case 1 [S_k a transient set] If S_k is a transient set, then S_i must be a transient state. Otherwise movement from S_i into S_k would be impossible. Theorem 3.1 now implies that $A^{\#}[i|S_k] > 0$ for this case.

Case 2 [S_k an ergodic set and S_i a transient state] In this case write the transition matrix as in (4.1) where T_{22} is the transition matrix associated with S_k, T_{33} corresponds to all other ergodic states, and T_{11} corresponds to the transient states. From (2.4) it follows that

$$A^{\#} = P' \left[\begin{array}{c|c|c} A_{11}^{-1} & A_{11}^{-2}A_{12}(I-A_{22}A_{22}^{\#})-A_{11}^{-1}A_{12}A_{22}^{\#} & A_{11}^{-2}A_{13}(I-A_{33}A_{33}^{\#}) -A_{11}^{-1}A_{13}A_{33}^{\#} \\ \hline 0 & A_{22}^{\#} & 0 \\ \hline 0 & 0 & A_{33}^{\#} \end{array}\right] P \tag{6.1}$$

State S_i being transient means that $A^{\#}[i|S_k]$ is equal to some row (say the r-th row) of the (1,2) - block of $PA^{\#}P'$. Let j = a vector of 1's and consider the sum of the entries in $A^{\#}[i|S_k]$ as follows.

$$(A^{\#}[i|S_k])j = e_r'[A_{11}^{-2}A_{12}(I-A_{22}A_{22}^{\#}) - A_{11}^{-1}A_{12}A_{22}^{\#}]j = -e_r'(I-T_{11})^{-2}T_{12}j \tag{6.2}$$

because $A_{22}j = A_{22}^{\#}j = 0$.

From (4.3) it is known that $(I-T_{11})^{-1}T_{12}j$ is the vector of absorption probabilities. That is,

$$(I-T_{11})^{-1}T_{12}j = \begin{bmatrix} p_1 \\ p_2 \\ \cdot \\ \cdot \\ p_n \end{bmatrix} \geq 0$$

where p_t = P (absorption into S_k | initially in the t-th transient state). State S_i is the r-th transient state so that the hypothesis that it is possible to eventually move from S_i into S_k implies that $p_r > 0$.

In the proof of Theorem 3.1 it was noted that the (r,s)-entry of $(I-T_{11})^{-1}$ gives the expected number of time spent in the state corresponding to row r when initially in the state corresponding to column s. The r-th row of $(I-T_{11})^{-1}$ is $e_r'(I-T_{11})^{-1} = (m_{r1}, m_{r2}, \ldots, m_{rn}) \geq 0$. Since $m_{rr} > 0$, it is now clear that $e_r'(I-T_{11})^{-2}T_{12}j > m_{rr}p_r > 0$ so that (6.2) implies $A^{\#}[i|S_k]j < 0$. Since the sum of the entries in $A^{\#}[i|S_k]$ is strictly negative, it must be the case that $A^{\#}[i|S_k] \neq 0$.

Case 3 [S_k an ergodic set and S_i an ergodic state] In this case, the fact that it is possible to reach S_k from S_i implies $S_i \in S_k = [S_i]$. As before, let T be written as in (4.1) so that $A^{\#}$ is given by (6.1). It is clear that $A^{\#}[i|S_k]$ is some row of $A_{22}^{\#}$. Lemma 6.1 now guarantees that $A^{\#}[i|S_k] \neq 0$. □

The previous theorem make it clear how to characterize the situation in which the process can eventually move from one irreducible class into another irreducible class.

Theorem 6.3 Let S_i and S_k be two irreducible classes of states. The class S_k is accessible from the class S_i if and only if $A^{\#}[S_i|S_k] \neq 0$.

Proof This is a direct consequence of Theorem 6.2. □

An Example

In order to illustrate the ideas of the previous sections and to obtain an idea of just how easy it really is to completely analyze a general chain once $A^{\#}$ is known, a non-trivial example is presented.

Consider a chain with 10 states whose transition matrix is given by

$$
T = \begin{bmatrix}
0.3 & 0.1 & 0 & 0 & 0 & 0 & 0.2 & 0 & 0.1 & 0.3 \\
0 & 0 & 0 & 0 & 0 & 0 & 0 & 0 & 1 & 0 \\
0 & 0.2 & 0 & 0 & 0.5 & 0 & 0 & 0 & 0 & 0.3 \\
0 & 0.5 & 0 & 0 & 0 & 0.5 & 0 & 0 & 0 & 0 \\
0.1 & 0 & 0.4 & 0 & 0.4 & 0.1 & 0 & 0 & 0 & 0 \\
0 & 0 & 0 & 0 & 0 & 0 & 0 & 0 & 1 & 0 \\
0 & 0 & 0 & 0 & 0 & 0 & 0.5 & 0.5 & 0 & 0 \\
0 & 0 & 0 & 0 & 0 & 0 & 1 & 0 & 0 & 0 \\
0 & 0 & 0 & 1 & 0 & 0 & 0 & 0 & 0 & 0 \\
0.5 & 0 & 0 & 0 & 0 & 0 & 0.3 & 0.2 & 0 & 0
\end{bmatrix}.
$$

To 5 significant digits, $A^{\#}$ and $I-AA^{\#}$ are computed to be

$$
A^{\#} = \begin{bmatrix}
1.8182 & -.035813 & 0 & -.31405 & 0 & -.21763 & -1.0002 & -.60312 & -.19284 & .54545 \\
0 & .66667 & 0 & -.33333 & 0 & -.33333 & 0 & 0 & 0 & 0 \\
0.63636 & -.12996 & 1.5 & -.8031 & 1.25 & -.3686 & -1.3904 & -.71624 & -.61901 & .64091 \\
0 & 0 & 0 & -.33333 & 0 & 0 & 0 & 0 & -.33333 & 0 \\
0.72727 & -.31357 & 0 & -.9741 & 2.5 & -.33629 & -1.5431 & -.80276 & -.77562 & .51818 \\
0 & -.33333 & 0 & -.33333 & 0 & .66667 & 0 & 0 & 0 & 0 \\
0 & 0 & 0 & 0 & 0 & 0 & .22222 & -.22222 & 0 & 0 \\
0 & 0 & 0 & 0 & 0 & 0 & -.44444 & .44444 & 0 & 0 \\
0 & -.16667 & 0 & 0 & 0 & .16667 & 0 & 0 & .33333 & 0 \\
0.90909 & -.048209 & 0 & -.21763 & 0 & -.13912 & -1.0678 & -.55207 & -.15702 & .2727
\end{bmatrix}
$$

$$
I-AA^{\#} = \begin{bmatrix}
0 & .060606 & 0 & .12121 & 0 & .060606 & .42424 & .21212 & .12121 & 0 \\
0 & .16667 & 0 & .33333 & 0 & .16667 & 0 & 0 & .33333 & 0 \\
0 & .092045 & 0 & .18409 & 0 & .092045 & .29848 & .14924 & .18409 & 0 \\
0 & .16667 & 0 & .33333 & 0 & .16667 & 0 & 0 & .33333 & 0 \\
0 & .099242 & 0 & .19848 & 0 & .099242 & .2697 & .13485 & .19848 & 0 \\
0 & .16667 & 0 & .33333 & 0 & .16667 & 0 & 0 & .33333 & 0 \\
0 & 0 & 0 & 0 & 0 & 0 & .66667 & .33333 & 0 & 0 \\
0 & 0 & 0 & 0 & 0 & 0 & .66667 & .33333 & 0 & 0 \\
0 & .16667 & 0 & .33333 & 0 & .16667 & 0 & 0 & .33333 & 0 \\
0 & .030303 & 0 & .060606 & 0 & .030303 & .54545 & .27273 & .060606 & 0
\end{bmatrix}
$$

By simply inspecting the elements of these matrices, one can reach the following conclusions.

1. The chain is not an ergodic chain. This is evident from Theorem 5.1 by noting that $I-AA^{\#}$ contains some zero entries.
2. $E(C) = 2$. That is, there are exactly two ergodic classes. This follows directly by computing Trace $(I-AA^{\#})$ and applying Theorem 5.5.

3. States S_1, S_3, S_5, S_{10} are each transient states while states $S_2, S_4, S_6, S_7, S_8, S_9$ are each ergodic states. This is deduced by inspecting either $I-AA^{\#}$ or $A^{\#}$ and using either Theorem 5.3 or Theorem 5.2.

4. By virtue of Theorem 5.4, or Corollary 5.1, it is clear from the entries of $A^{\#}$ that $\{S_1, S_{10}\}$ is one transient class while $\{S_3, S_5\}$ is another transient class.

5. The two ergodic classes are obtained by applying Theorem 5.6 to $I-AA^{\#}$. They are easily seen to be $E_1 = \{S_2, S_4, S_6, S_9\}$ and $E_2 = \{S_7, S_8\}$. (one could have also used Theorem 5.7 to determine these ergodic sets).

6. The mean absorption times from each transient state are available from $A^{\#}$ via Theorem 3.2. They are given in the following table.

Initial State (S_i)	Mean Absorption Time ($\sum_{k=1,3,5,10} A^{\#}[i\|k]$)
S_1	2.3636
S_3	4.0273
S_5	4.7455
S_{10}	2.1818

7. The probability of absorption into each of the two ergodic classes is obtained from Theorem 4.2 by inspecting the entries of $I-AA^{\#}$. ($\sum (I-AA^{\#})[i|k]$)

Matrix of Absorption Probabilities

	E_1	E_2
S_1	.36363	.63637
S_3	.55227	.44773
S_5	.59545	.40455
S_{10}	.18182	.81818

8. The eventual flow of the process, as shown below is evident by applying either Theorem 6.2 or Theorem 6.3 to $A^{\#}$. (A dotted arrow - - - - - → indicates the possibility of <u>eventual</u> movement.

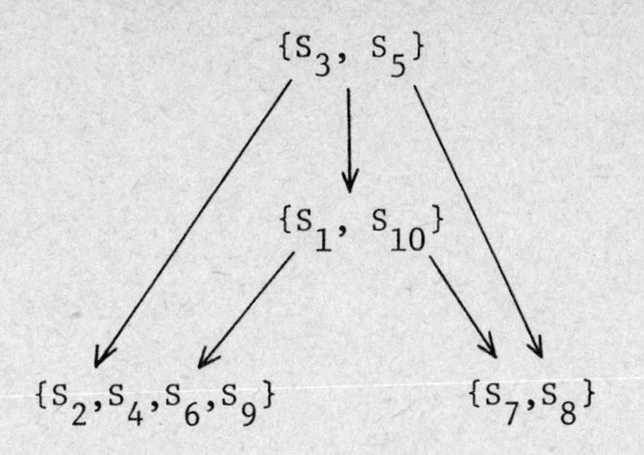

One of the largest advantages of using $A^{\#}$ over the traditional **techniques** is that the transition matrix T need never be reduced to a canonical form such as that of (2.6). All of the information and characteristics of the chain can be deduced directly from either $A^{\#}$ or $I-AA^{\#}$. Moreover, it is not necessary to have T in canonical form in order to compute $A^{\#}$ and $I-AA^{\#}$. In fact, once $A^{\#}$ is known, the canonical form for T is obvious. In the above example, it is clear from part 8 that rearranging the states in the order $S_3, S_5, S_1, S_{10}, S_2, S_4, S_6, S_9, S_7, S_8$ produces the transition matrix.

	3	5	1	10	2	4	6	9	7	8
3	0	.5	0	.3	.2	0	0	0	0	0
5	.4	.4	.1	0	0	0	.1	0	0	0
1	0	0	.3	.3	.1	0	0	.1	.2	0
10	0	0	.5	0	0	0	0	0	.3	.2
2	0	0	0	0	0	0	0	1	0	0
4	0	0	0	0	.5	0	.5	0	0	0
6	0	0	0	0	0	0	0	1	0	0
9	0	0	0	0	0	1	0	0	0	0
7	0	0	0	0	0	0	0	0	.5	.5
8	0	0	0	0	0	0	0	0	1	0

which is the canonical form given in Lemma 2.3.

The following chart summarizes the use of $A^{\#}$ in analyzing a general finite Markov chain.

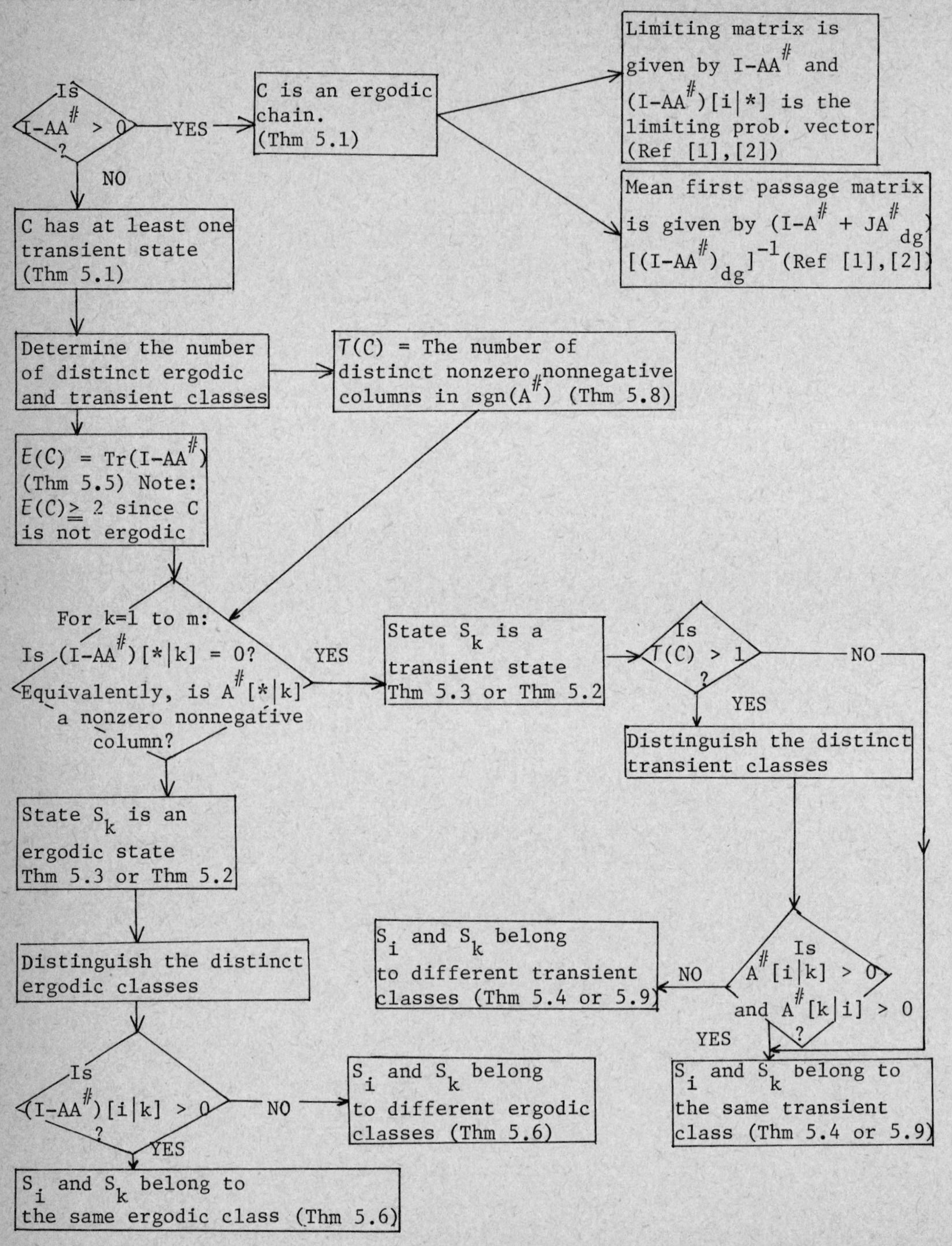

REFERENCES

1 S. L. Campbell and Carl D. Meyer, Jr., Generalized Inverses of Linear Transformations, Surveys and Reference Works in Mathematics, Pitman Pub. Co., London, 1979.

2 Carl D. Meyer, Jr., The role of the group generalized inverse in the theory of finite Markov chains, SIAM Rev., 17(1975), pp. 443-464.

3 Carl D. Meyer, Jr., The condition of a finite Markov chain and perturbation bounds for the limiting probabilities, SIAM J. Alg. Disc. Meth., 1(1980), pp. 273-283.

4 Carl D. Meyer, Jr., and N. J. Rose, The index and the Drazin inverse of block triangular matrices, SIAM J. Appl. Math., 33(1977), pp. 1-7.

Carl D. Meyer, Jr.
Mathematics Department
North Carolina State University
Raleigh, NC 27650

J H WILKINSON

Note on the practical significance of the Drazin inverse

ABSTRACT

The solution of the differential system $B\dot{x} = Ax + f$ where A and B are $n \times n$ matrices and $A - \lambda B$ is not a singular pencil may be expressed in terms of the Drazin inverse. It is shown that there is a simple reduced form for the pencil $A - \lambda B$ which is adequate for the determination of the general solution and that although the Drazin inverse could be determined efficiently from this reduced form it is inadvisable to do so.

1. INTRODUCTION

In a recent paper [2] the solution of the differential system

$$B\dot{x} = Ax + f(t), \tag{1.1}$$

where B and A are $n \times n$ matrices and f is an n-vector has been discussed in terms of the Drazin inverse. Although this work gives considerable insight into the nature of the general solution of (1.1) it should not be assumed that because the explicit solution can be expressed directly in terms of the Drazin inverse that economical algorithms will involve its explicit computation.

Numerical analysts will be familiar with this in connection with the simpler problem $Ax = b$ where A is non-singular. Although the solution is given by $x = A^{-1}b$ it is seldom advisable to compute the inverse explicitly. However algorithms for solving $Ax = b$ based on direct methods do provide the basic tools for the efficient computation of A^{-1} if that should be required; we might therefore expect that practical algorithms for solving (1.1), or closely related algorithms, would provide effective methods for computing the Drazin inverse and this is indeed true.

2. THE DRAZIN INVERSE

If A is an $n \times n$ matrix, then the Drazin inverse [4] of A is the matrix X satisfying the relations

(i) $AX = XA$

(ii) $XAX = X$

(iii) $XA^{k+1} = A^k$, where $k = \text{Ind}(A)$.

Ind(A), the index of A, is the smallest non-negative integer for which rank (A^k) = rank (A^{k+1}).

The existence and uniqueness of A may be proved as follows. The proof is given in matrix terms since we shall need to work in these terms in subsequent sections. Let J be the Jordan canonical form of A, and suppose J is expressed as the direct sum of C and N where C is associated with the non-zero eigenvalues and N is associated with the zero eigenvalues and is therefore nil-potent. We may write

$$A = T \begin{bmatrix} C & 0 \\ 0 & N \end{bmatrix} T^{-1}, \tag{2.1}$$

where C is non-singular and N is nil-potent. If k is the smallest integer for which $N^k = 0$, it is clear that k is the index of A since

$$A^k = T \begin{bmatrix} C^k & 0 \\ 0 & 0 \end{bmatrix} T^{-1}, \quad A^{k+1} = T \begin{bmatrix} C^{k+1} & 0 \\ 0 & 0 \end{bmatrix} T^{-1} \tag{2.2}$$

and $\text{rank}(A^k) = \text{rank}(A^{k+1})$ = order of C. On the other hand $\text{rank}(A^p) >$ $\text{rank}(A^{p+1})$ when $p < k$. Obviously k is the dimension of the largest Jordan submatrix associated with a zero eigenvalue.

Any $n \times n$ matrix X may be expressed in the form $X = TYT^{-1}$ and relations (i), (ii) and (iii) are satisfied if and only if

(iv) $JY = YJ$,

(v) $YJY = Y$,

(vi) $YJ^{k+1} = J^k$,

where

$$J = \begin{bmatrix} C & 0 \\ 0 & N \end{bmatrix}. \tag{2.3}$$

Partitioning Y conformally with J we may write

$$Y = \begin{bmatrix} P & Q \\ R & S \end{bmatrix} . \tag{2.4}$$

Equation (iv) then gives

$$\begin{aligned} CP &= PC \ (a) , \quad CQ = QN \ (b) \\ NR &= RC \ (c) , \quad NS = SN \ (d) \end{aligned} . \tag{2.5}$$

From (b) we have

$$CQN^{k-1} = QN^k = 0. \tag{2.6}$$

Hence $QN^{k-1} = 0$ since C is non-singular. Continuing in this way we have successively $QN^{k-2} = 0$, $QN^{k-3} = 0$, ..., $Q = 0$. Similarly from (c), $R = 0$. Now from (v) and (d)

$$SNS = S \text{ and } S^2N = S. \tag{2.7}$$

Hence

$$S^2N^k = SN^{k-1} \text{ giving } SN^{k-1} = 0. \tag{2.8}$$

Continuing in this way $SN^{k-2} = 0$, SN^{k-3}, ..., $S = 0$. Finally from (vi),

$$PC^{k+1} = C^k \text{ giving } P = C^{-1} \tag{2.9}$$

and hence

$$X = T \begin{bmatrix} C^{-1} & 0 \\ 0 & 0 \end{bmatrix} T^{-1} , \tag{2.10}$$

showing that X is uniquely determined. In proving this result we did not make use of the fact that C and N were the direct sum of Jordan matrices but merely that they were non-singular and nil-potent respectively. Hence to derive the Drazin inverse it is not necessary to obtain the Jordan canonical form itself but merely the identification of the nil-potent part, a much simpler objective.

When A is non-singular X is obviously A^{-1}, the usual inverse. Notice that it is not generally true that $AXA = A$ and hence a solution of a compatible system $Ax = b$ is not, in general, given by $x = Xb$.

3. COMPUTATION OF THE DRAZIN INVERSE

We have shown that the Drazin inverse of A is available if we have expressed A in the form

$$A = T \left[\begin{array}{c|c} C & 0 \\ \hline 0 & N \end{array}\right] T^{-1} , \qquad (3.1)$$

where C is non-singular and N is nil-potent. A factorization of that form in which T is unitary has in fact been derived by Golub and Wilkinson [6]. In that factorization the singular value decomposition was used so as to give the maximum numerical stability. A similar reduction could be achieved by a whole range of elementary transformations and this we now describe in general terms.

We denote the original matrix by $A^{(1)}$. In the rth step a similarity transformation, based on multiplications with elementary matrices, is applied to $A^{(r)}$ to give $A^{(r+1)}$. The general form of the matrices $A^{(r)}$ is adequately illustrated by the fact that

$$A^{(4)} = \left[\begin{array}{c|c|c|c} A_{44}^{(4)} & A_{43}^{(4)} & A_{42}^{(4)} & A_{41}^{(4)} \\ \hline 0 & 0 & A_{32}^{(4)} & A_{31}^{(4)} \\ \hline 0 & 0 & 0 & A_{21}^{(4)} \\ \hline 0 & \underbrace{0}_{n_3} & \underbrace{0}_{n_2} & \underbrace{0}_{n_1} \end{array}\right] \begin{array}{l} \\ \}\, n_3 \\ \}\, n_2 \\ \}\, n_1 \end{array} , \qquad (3.2)$$

where the significance of the n_i will become apparent in the description of the rth step which is as follows.

If the matrix $A_{rr}^{(r)}$ is non-singular, the reduction is complete. Otherwise premultiply $A_{rr}^{(r)}$ with a sequence of elementary transformations, the product of which is denoted by $Q^{(r)}$, such that

$$Q^{(r)} A_{rr}^{(r)} = \left[\begin{array}{c} B^{(r)} \\ \hline 0 \end{array}\right] \Big\} n_r \qquad , \tag{3.3}$$

where n_r is the nullity of $A_{rr}^{(r)}$. The matrices involved in $Q^{(r)}$ may be unitary (orthogonal, if real) or may be elementary matrices corresponding to elimination techniques. If $A^{(1)}$ had small integer elements the use of rational numbers enables this reduction to be done exactly. Note that $B^{(r)}$ need not be trapezoidal so that this reduction can be achieved entirely by pre-multiplications. If we now post-multiply by $(Q^{(r)})^{-1}$ we may write

$$Q^{(r)} A_{rr}^{(r)} (Q^{(r)})^{-1} = \left[\begin{array}{c|c} A_{r+1,r+1}^{(r+1)} & A_{r+1,r}^{(r+1)} \\ \hline 0 & 0 \end{array}\right]. \tag{3.4}$$

Writing

$$T^{(r)} = \left[\begin{array}{c|c} Q^{(r)} & 0 \\ \hline 0 & I \end{array}\right] \qquad , \tag{3.5}$$

where $T^{(r)}$ is of order n, then $A^{(r+1)} = T^{(r)} A^{(r)} (T^{(r)})^{-1}$ is again of the required form. Notice that the pre-multiplication with $T^{(r)}$ affects only the leading block row of $A^{(r)}$, while the post-multiplication affects only the principal leading submatrix. We must have $n_r \leq n_{r-1}$ since if $n_r > n_{r-1}$, this would imply that in the preceding stage n_{r-1} was not the nullity. Indeed the $A_{i+1,i}^{(k)}$ must be of full row rank at every stage for the same reason.

If the matrix $A^{(1)}$ is entirely nil-potent, then we must reach an $A_{kk}^{(k)}$ which is null and the final matrix is of the block form illustrated by

$$\begin{bmatrix} 0 & X & X & X \\ 0 & 0 & X & X \\ 0 & 0 & 0 & X \\ 0 & 0 & 0 & 0 \end{bmatrix} . \tag{3.6}$$

Otherwise we terminate with an $A^{(k+1)}_{k+1,k+1}$ which is non-singular. (In using the symbol k we are anticipating that this is the index of $A^{(1)}$). In this second case we can annihilate all blocks in the first row except $A^{(k+1)}_{k+1,k+1}$ by further similarity transformations. This is adequately illustrated by the case when $k = 3$ for which $A^{(4)}$ is as in (3.2) with $A^{(4)}_{44}$ non-singular. Post-multiplication with

$$P_3 = \begin{bmatrix} I & -(A^{(4)}_{44})^{-1}A^{(4)}_{43} & \cdot & 0 \\ 0 & I & & \cdot \\ \cdot & & I & \cdot \\ 0 & \cdot & \cdot & I \end{bmatrix} \tag{3.7}$$

annihilates $A^{(4)}_{43}$ and leaves all other submatrices unaltered. Pre-multiplication with $P_3^{(-1)}$ preserves all the null matrices and changes $A^{(4)}_{42}$ and $A^{(1)}_{41}$. The (4,2) and (4,1) blocks may be annihilated successively in a similar way.

Thus according as $A^{(1)}$ is entirely nil-potent or not we achieve a reduction to one or other of the forms illustrated by

$$\begin{bmatrix} 0 & X & X & X \\ 0 & 0 & X & X \\ 0 & 0 & 0 & X \\ 0 & 0 & 0 & 0 \end{bmatrix} \text{ or } \begin{bmatrix} C & 0 & 0 & 0 \\ 0 & 0 & X & X \\ 0 & 0 & 0 & X \\ 0 & 0 & 0 & 0 \end{bmatrix} \tag{3.8}$$

with C non-singular. We may denote this final matrix by

$$N \text{ or } \begin{bmatrix} C & 0 \\ 0 & N \end{bmatrix} \tag{3.9}$$

in the two cases. Obviously $N^k = 0$ while it is easy to see that since the $(i,i+1)$ blocks are all of full row rank $N^\ell \neq 0 (\ell < k)$. Hence k is indeed the index.

The Drazin inverse could now be computed explicitly using the product of all the transformation matrices but it would usually be more expedient to keep it in factorized form.

4. THE SOLUTION OF THE DIFFERENTIAL SYSTEM

When B is non-singular, the system (1.1) may be written in the form

$$\dot{x} = B^{-1}Ax + B^{-1}f \; . \tag{4.1}$$

There is a solution corresponding to any f and for arbitrary initial values x_o. This solution may be expressed in terms of $\exp(B^{-1}At)$. Singularity of A in no way affects the explicit form of the solution. Although this is a non-trivial matter we shall assume, in common with the paper we have referred to, that we have satisfactory algorithms for it.

When B is singular but A is non-singular, (1.1) may be written in the form

$$A^{-1}B\dot{x} = x + A^{-1}f \tag{4.2}$$

ie

$$K\dot{x} = x + g \text{ (say)} \quad . \tag{4.3}$$

The existence and nature of the solution may be examined in terms of the Drazin inverse of K, but there seems to be little point in computing the latter explicitly. Indeed if

$$K = T^{-1}\begin{bmatrix} C & 0 \\ 0 & N \end{bmatrix} T, \tag{4.4}$$

then

$$\begin{bmatrix} C & 0 \\ 0 & N \end{bmatrix} T\dot{x} = Tx + Tg, \tag{4.5}$$

or

$$\begin{bmatrix} C & 0 \\ 0 & N \end{bmatrix}\begin{bmatrix} \dot{y} \\ \dot{z} \end{bmatrix} = \begin{bmatrix} y \\ z \end{bmatrix} + \begin{bmatrix} p \\ q \end{bmatrix}, \tag{4.6}$$

$$\begin{bmatrix} y \\ z \end{bmatrix} = Tx \quad , \quad \begin{bmatrix} p \\ q \end{bmatrix} = Tg \quad . \tag{4.7}$$

Hence

$$C\dot{y} = y + p \tag{4.8}$$

$$N\dot{z} = z + q \quad . \tag{4.9}$$

Since $N^k = 0$, (4.9) gives

$$0 = N^{k-1}z + N^{k-1}\dot{q} \quad . \tag{4.10}$$

Multiplying (4.9) by N^{k-2} and substituting from (4.10);

$$-N^{k-1}\dot{q} = N^{k-2}z + N^{k-2}q \tag{4.11}$$

and continuing in this way

$$z = -[I + ND + \ldots + N^{k-1}D^{k-1}]\, q \quad \text{where} \quad D = \frac{d}{dt} \quad . \tag{4.12}$$

Notice that we must have

$$z_o = (-[I + ND + \ldots + N^{k-1}D^{k-1}]\, q)_o \tag{4.13}$$

and since the components of z_o are linear combinations of those of x_o this means that the initial x_o must satisfy certain conditions for a solution to be possible. Provided these consistency conditions are satisfied there is then a unique solution corresponding to any q, assuming that it has $k-1$ derivatives. We observe that in the homogeneous case $q = 0$, and the only solution of (4.9) is $z = 0$.

Since C is non-singular the system (4.8) has a unique solution corresponding to any initial y_o and this may be expressed in terms of $\exp(C^{-1}t)$.

The solution described above has been given in the spirit of the work based on the use of the Drazin inverse, but we would submit that even here too much attention has been paid to obtaining explicit expressions. It is more economical to work with the form exemplified in (3.2). We describe this below and for convenience of presentation we assume that $k = 3$ and omit upper suffices. A transformation of variables has then reduced the original system to one of the form

$$\begin{bmatrix} A_{44} & A_{43} & A_{42} & A_{41} \\ 0 & 0 & A_{32} & A_{31} \\ 0 & 0 & 0 & A_{21} \\ 0 & 0 & 0 & 0 \end{bmatrix} \begin{bmatrix} \dot{y}_4 \\ \dot{y}_3 \\ \dot{y}_2 \\ \dot{y}_1 \end{bmatrix} = \begin{bmatrix} y_4 \\ y_3 \\ y_2 \\ y_1 \end{bmatrix} + \begin{bmatrix} g_4 \\ g_3 \\ g_2 \\ g_1 \end{bmatrix} \tag{4.14}$$

where the blocks on the diagonal are square and A_{44} is non-singular. The matrix

$$\begin{bmatrix} 0 & A_{32} & A_{31} \\ 0 & 0 & A_{21} \\ 0 & 0 & 0 \end{bmatrix} \tag{4.15}$$

is the N and A_{44} is the C of our previous analysis.

The relation (4.14) gives successively

$$y_1 = -g_1, \quad y_2 = -g_2 - A_{21}\dot{y}_1, \quad y_3 = -g_3 - A_{31}\dot{y}_1 - A_{32}\dot{y}_2 \quad . \tag{4.16}$$

Finally we have

$$A_{44}\dot{y}_4 = y_4 + (g_4 - A_{41}\dot{y}_1 - A_{42}\dot{y}_2 - A_{43}\dot{y}_3) \tag{4.17}$$

and at this stage y_1, y_2 and y_3 and hence $\dot{y}_1$, $\dot{y}_2$ and $\dot{y}_3$ have already been determined. Notice that when we describe the solution in these terms there is no need to annihilate the blocks A_{43}, A_{42} and A_{41} as we did in section 3 when describing a reduction to the form

$$\begin{bmatrix} C & 0 \\ 0 & N \end{bmatrix} . \tag{4.18}$$

Now we merely have terms involving these A_{4i} on the right of (4.17). At the end of the next section we show how the volume of work may be reduced even further.

5. SINGULAR A AND B

When both A and B are singular one cannot proceed as in the previous section. The use of the Drazin inverse has been concerned with the case when $\det(A-\lambda B) \not\equiv 0$ ie when the pencil $A-\lambda B$ is non-singular in the Kronecker sense (see eg [3,5,8]). The matrix $A-cB$ is then non-singular for any c which is not a root of the equation $\det(A-\lambda B) = 0$. If one takes any such c then the system (1.1) is equivalent to

$$(A-cB)^{-1}B\dot{x} = (A-cB)^{-1}Ax + (A-cB)^{-1}f \tag{5.1}$$

or

$$\hat{B}\dot{x} = \hat{A}x + f \quad . \tag{5.2}$$

It may be readily verified that $\hat{B}\hat{A} = \hat{A}\hat{B}$. The explicit solution of (5.2) may be expressed in terms of the Drazin inverse of $\hat{B}$. Although, of course, the derived solution must be independent of c, its introduction is undesirable. In practice it would be important for $A-cB$ to be, not merely non-singular, but well conditioned with respect to inversion, otherwise there will be a loss of accuracy which may be far greater than that resulting from the inherent sensitivity of the problem.

It will be appreciated that one will not necessarily know in advance whether A and B are singular or indeed whether $\det(A-\lambda B) \equiv 0$. The method described below, which is analogous to that described in section 3 for the computation of the Drazin inverse of a matrix, does not require any previous knowledge and does not require the use of the arbitrary scalar c.

We observe that if P and Q are non-singular, then pre-multiplication of the system (1.1) with P and the transformation $x = Qy$ transforms it to the equivalent system

$$PBQ\dot{y} = PAQy + Pf \quad . \tag{5.3}$$

In our algorithm P and Q are determined as products of elementary matrices in such a way that (5.3) is typically of the form illustrated by

$$\begin{bmatrix} B_{44}^{(4)} & B_{43}^{(4)} & B_{42}^{(4)} & B_{41}^{(4)} \\ 0 & 0 & B_{32}^{(4)} & B_{31}^{(4)} \\ 0 & 0 & 0 & B_{21}^{(4)} \\ 0 & 0 & 0 & 0 \end{bmatrix} \begin{bmatrix} \dot{y}_4 \\ \dot{y}_3 \\ \dot{y}_2 \\ \dot{y}_1 \end{bmatrix} = \begin{bmatrix} A_{44}^{(4)} & A_{43}^{(4)} & A_{42}^{(4)} & A_{41}^{(4)} \\ 0 & A_{33}^{(4)} & A_{32}^{(4)} & A_{31}^{(4)} \\ 0 & 0 & A_{22}^{(4)} & A_{21}^{(4)} \\ 0 & 0 & 0 & A_{11}^{(4)} \end{bmatrix} \begin{bmatrix} y_4 \\ y_3 \\ y_2 \\ y_1 \end{bmatrix} + \begin{bmatrix} g_4 \\ g_3 \\ g_2 \\ g_1 \end{bmatrix} . \tag{5.4}$$

The diagonal blocks are square and $A_{11}^{(4)}$, $A_{22}^{(4)}$, $A_{33}^{(4)}$, and $B_{44}^{(4)}$ are non-singular. The matrices $B_{21}^{(4)}$ and $B_{32}^{(4)}$ are of full row rank. In general there are k steps, the process coming to an end when $B_{k+1,k+1}^{(k+1)}$ is non-singular.

Suppose we have performed $r-1$ steps and $B_{r,r}^{(r)}$ is still singular. In this case $B_{r,r}^{(r)}$ may be reduced to the form

$$\left[\begin{array}{c} E^{(r)} \\ \hline 0 \end{array}\right] \} \; n_r \tag{5.5}$$

by pre-multiplication with elementary matrices. Here n_r is the nullity of $B_{r,r}^{(r)}$ and $E^{(r)}$ is not required to be of upper trapezoidal form. If the same operations are performed on $A_{rr}^{(r)}$ the resulting matrix may be denoted by

$$\left[\begin{array}{c} F^{(r)} \\ \hline G^{(r)} \end{array}\right] \} \; n_r \quad . \tag{5.6}$$

Now $G^{(r)}$ must be of full row rank n_r, since otherwise $A_{rr}^{(r)}$ and $B_{rr}^{(r)}$ share a common left-hand null vector and this would imply that $\det(A_{rr}^{(r)} - \lambda B_{rr}^{(r)}) \equiv 0$. Hence $G^{(r)}$ may be multiplied on the right by elementary matrices to give

$$[0 \mid A_{rr}^{(r+1)} \;] \; , \tag{5.7}$$

where $A_{rr}^{(r+1)}$ is non-singular. If these right-hand transformations are applied to the full matrices,

$$\left[\begin{array}{c} E^{(r)} \\ \hline 0 \end{array}\right] \text{ and } \left[\begin{array}{c} F^{(r)} \\ \hline G^{(r)} \end{array}\right] \quad , \tag{5.8}$$

the resulting matrices may be denoted by

$$\left[\begin{array}{c|c} B^{(r+1)}_{r+1,r+1} & B^{(r+1)}_{r+1,r} \\ \hline 0 & 0 \end{array}\right] \quad \text{and} \quad \left[\begin{array}{c|c} A^{(r+1)}_{r+1,r+1} & A^{(r+1)}_{r+1,r} \\ \hline 0 & A^{(r+1)}_{rr} \end{array}\right]. \tag{5.9}$$

The $r\underline{th}$ step is completely determined by the matrices $B^{(r)}_{rr}$ and $A^{(r)}_{rr}$ but if we apply the transformations to the full $n \times n$ matrices and to the current forcing vector we arrive at an $r\underline{th}$ derived system of the same form as the $(r-1)$th system from which we started. The $B^{(r+1)}_{r,r-1}$ must be of full row rank otherwise the n_{r-1} determined in the previous stage would have been incorrect.

If $\det(A-\lambda B) \not\equiv 0$ we must either reach a $B^{(r)}_{rr}$ which is non-singular or one which is completely null with $A^{(r)}_{rr}$ non-singular. If however $\det(A-\lambda B) \equiv 0$ this would be detected by the algorithm since we would reach a stage at which the $G^{(r)}$ of (5.6) was not of full rank and this would reveal itself when performing the elementary operations on $G^{(r)}$.

For simplicity of presentation let us assume that the process terminates when $k = 3$ so that the final system is as given in (5.4). We suppress the upper suffix for convenience. The solution is then given by

$$\begin{aligned} A_{11}y_1 &= -g_1 \\ A_{22}y_2 &= -g_2 - A_{21}y_1 - B_{21}\dot{y}_1 \\ A_{33}y_3 &= -g_3 - A_{31}y_1 - A_{32}y_2 - B_{31}\dot{y}_1 - B_{32}\dot{y}_2 \end{aligned} \tag{5.10}$$

so that the components of y_1, y_2 and y_3 are all uniquely determined and the initial values must satisfy equations (5.10) for consistency. Finally

$$B_{44}\dot{y}_4 = A_{44}y_4 + (A_{43}y_3 + A_{42}y_2 + A_{41}y_1 - B_{43}\dot{y}_3 - B_{42}\dot{y}_2 - B_{41}\dot{y}_1 + g_4) \tag{5.11}$$

and the vector in parenthesis is already determined. Since B_{44} is non-singular this has a unique solution for arbitrary initial y_4 which may be expressed in terms of $\exp(B^{-1}_{44} A_{44}t)$.

The elementary transformations on $G^{(r)}$ would usually be carried out in such a way that $A^{(r+1)}_{rr}$ would be at least triangular (though possibly even diagonal) according to the method used. The computation of the vectors y_1, y_2, y_3 from relations (5.10) would therefore be particularly convenient. As we remarked above if at any stage $G^{(r)}$ is not of full rank this would be exposed automatically in the execution of the algorithm. (We assume here that the algorithm used to reduce $G^{(r)}$ is stable enough to detect rank reliably!). This can happen only if $\det(A-\lambda B) \equiv 0$. This situation is not usually covered by the use of the Drazin inverse. When $G^{(r)}$ has a rank deficiency of p, then p linear relations must hold between components of f for the differential equations to be compatible. This is discussed in detail in [8]. However the general situation may be illustrated by considering what happens when $G^{(1)}$ has a rank deficiency of p. This means that the original system is equivalent to a system of the form

$$\begin{matrix} n_1 - p \\ \\ p \end{matrix}\left\{\begin{bmatrix} K \\ \hline 0 \\ \hline 0 \end{bmatrix}\right. \dot{y} = \begin{matrix} n_1 - p \\ \\ p \end{matrix}\left\{\begin{bmatrix} L \\ \hline M \\ \hline 0 \end{bmatrix}\right. y + g , \qquad (5.12)$$

where M is of full rank, n_1-p. Hence the last p components of g must be zero for the equations to be compatible, and the components of g are linear combinations of the original components of f.

When both A and B are singular but $\det(A-\lambda B) \not\equiv 0$, then when we reach the terminating non-singular $B_{rr}^{(r)}$ the corresponding $A_{rr}^{(r)}$ must be singular. This follows because the earlier $A_{ii}^{(i)}$ were non-singular and if $A_{rr}^{(r)}$ were non-singular this would imply non-singularity of A.

We have remarked that the solution may be expressed in terms of the Drazin inverse of $(A-cB)^{-1}B$ and the form of the solution is determined by the index of $(A-cB)^{-1}B$. The k introduced above is in fact this index as we now show. Denoting the successive $n \times n$ matrices derived by the algorithm by $A^{(r)}$ and $B^{(r)}$ respectively, $A^{(k+1)} - cB^{(k+1)}$ has as its diagonal blocks

$$A_{k+1,k+1}^{(k+1)} - cB_{k+1,k+1}^{(k+1)} , \quad A_{kk}^{(k+1)} , \ldots , \quad A_{2,2}^{(k+1)} , \quad A_{11}^{(k+1)} . \qquad (5.13)$$

The last k of these and $B_{k+1,k+1}^{(k+1)}$ are non-singular by definition of the algorithm. The first is non-singular for any c for which

$\det(A^{(k+1)}_{k+1,k+1} - cB^{(k+1)}_{k+1,k+1}) \neq 0$ ie for almost all c. Obviously

$$X = [A^{(k+1)} - cB^{(k+1)}]^{-1} \quad B^{(k+1)}$$

is block upper-triangular and its diagonal blocks are

$$[A^{(k+1)}_{k+1,k+1} - cB^{(k+1)}_{k+1,k+1}]^{-1} \quad B^{(k+1)}_{k+1,k+1}, 0, \ldots, 0, 0 . \qquad (5.14)$$

Further $X_{i,i-1} = (A^{(k+1)}_{ii})^{-1} \; B^{(k+1)}_{i,i-1}$ and hence is of full row rank for $2 \leq i \leq k$ since this is true of the $B^{(k+1)}_{i,i-1}$. Hence the k of our algorithm is the index of $[A^{(k+1)} - cB^{(k+1)}]^{-1} B^{(k+1)}$ and since $A^{(k+1)} = PAQ$, $B^{(k+1)} = PBQ$ for some non-singular P and Q, our k is the index of $(A-cB)^{-1}B$.

The algorithm we have described works in terms of full $n \times n$ matrices at all stages in the reduction, though to be sure in later stages only parts of these matrices are affected by the transformation. We have presented the algorithm in this way in order to give a closer tie up with earlier work involving the Drazin inverse. However, if one were concerned with only one forcing vector f, or if indeed one were interested in several different forcing functions all of which were known at the time when the reduction was performed, then a considerable economy would be achieved as follows. Suppose we have completed one stage of the reduction and have reached the reduced system

$$\begin{bmatrix} B^{(2)}_{22} & B^{(2)}_{21} \\ 0 & 0 \end{bmatrix} \begin{bmatrix} \dot{y}_2 \\ \dot{y}_1 \end{bmatrix} = \begin{bmatrix} A^{(2)}_{22} & A^{(2)}_{21} \\ 0 & A^{(2)}_{11} \end{bmatrix} \begin{bmatrix} y_2 \\ y_1 \end{bmatrix} + \begin{bmatrix} g_2 \\ g_1 \end{bmatrix} . \qquad (5.15)$$

At this stage the variables in y_1 are completely determined and these variables undergo no further transformations. We have then

$$y_1 = - (A^{(2)}_{11})^{-1} \quad g_1 \qquad (5.16)$$

and

$$B_{22}^{(2)}\dot{y}_2 = A_{22}^{(2)} y_2 + \{g_2 - B_{21}^{(2)}\dot{y}_1 + A_{21}^{(2)} y_1\}$$

$$= A_{22}^{(2)} y_2 + f_2 \text{ (say)} \quad . \tag{5.17}$$

Hence we can continue with a system of lower order. In this way we avoid performing any transformations on $B_{21}^{(2)}$ and $A_{21}^{(2)}$ in the next step. The first stage is wholly typical; in the rth stage we determine n_r more variables and are left with a system in n_r fewer variables. Obviously if we are interested in the effect of several forcing functions we can deal with them all simultaneously. A similar reduction of effort may be achieved with the simpler algorithm of section 4.

6. NUMERICAL EXAMPLE

As an illustration of our algorithm we describe its performance on the example used by Campbell et al [2]. The system of differential equations is

$$A\dot{x} + Bx = b$$

$$\begin{bmatrix} -1 & 0 & 2 \\ 2 & 3 & 2 \\ 1 & 0 & -2 \end{bmatrix} \dot{x} + \begin{bmatrix} -27 & -22 & -17 \\ 18 & 14 & 10 \\ 0 & 1 & 2 \end{bmatrix} x = \begin{bmatrix} 2 \\ 0 \\ 1 \end{bmatrix} , \tag{6.1}$$

where we have reordered the equations in order to avoid a row permutation during the course of the solution. This makes the process a little easier to follow. Naturally we have used rational elimination techniques. The authors gave the general solution to the homogeneous system as well as that corresponding to the forcing function b. For convenience of comparison we have followed the notation $A\dot{x} + Bx = b$ used by Campbell et al.

Exposing the row nullity of A gives

$$\begin{bmatrix} -1 & 0 & 2 \\ 2 & 3 & 2 \\ 0 & 0 & 0 \end{bmatrix} \dot{x} + \begin{bmatrix} -27 & -22 & -17 \\ 18 & 14 & 10 \\ -27 & -21 & -15 \end{bmatrix} x = \begin{bmatrix} 2 \\ 0 \\ 3 \end{bmatrix} . \tag{6.2}$$

We now reduce the rows of B corresponding to the null rows of A. In fact there is only one such row and to facilitate comparison with Campbell et al we leave (3,1) as the non-zero element rather than (3,3). This involves the transformation

$$x = \begin{bmatrix} 1 & -\frac{7}{9} & -\frac{5}{9} \\ 0 & 1 & 0 \\ 0 & 0 & 1 \end{bmatrix} y \text{ or } \begin{aligned} y_1 &= x_1 + \tfrac{7}{9}x_2 + \tfrac{5}{9}x_3 \\ y_2 &= x_2 \\ y_3 &= x_3 \end{aligned} \tag{6.3}$$

and leads to

$$\begin{bmatrix} -1 & \frac{7}{9} & \frac{23}{9} \\ 2 & \frac{13}{9} & \frac{8}{9} \\ 0 & 0 & 0 \end{bmatrix} \dot{y} + \begin{bmatrix} -27 & -1 & -2 \\ 18 & 0 & 0 \\ -27 & 0 & 0 \end{bmatrix} y = \begin{bmatrix} 2 \\ 0 \\ 3 \end{bmatrix} . \tag{6.4}$$

At this stage the singularity of B is exposed. The third equation gives

$$-27y_1 = 3 \quad \text{ie} \quad 9x_1 + 7x_2 + 5x_3 + 1 = 0 \tag{6.5}$$

while for the homogeneous system

$$9x_1 + 7x_2 + 5x_3 = 0 \quad . \tag{6.6}$$

Notice that these relations must hold for all values of t and therefore in particular for $t = 0$; at $t = 0$ they are in fact equations (35) and (29) respectively of Campbell et al.

Substituting $y_1 = -1/9$ into the first two equations and remembering that $y_2 = x_2$, $y_3 = x_3$ we have

$$\begin{aligned} \frac{7}{9}\dot{x}_2 + \frac{23}{9}\dot{x}_3 - x_2 - 2x_3 &= -1 \\ \frac{13}{9}\dot{x}_2 + \frac{8}{9}\dot{x}_3 &= 2 \end{aligned} \tag{6.7}$$

and the solution is now trivial. The general solution is

$$x_1 = -\frac{1}{18}(x_2(0) + 2x_3(0))e^{2/3t} - \frac{1}{18}(13x_2(0) + 8x_3(0)) - t - \frac{1}{9}$$

$$x_2 = -\frac{1}{18}(8x_2(0) + 16x_3(0))e^{2/3t} + \frac{1}{18}(26x_2(0) + 16x_3(0)) + 2t \qquad (6.8)$$

$$x_3 = \frac{1}{18}(13x_2(0) + 26x_3(0))e^{2/3t} - \frac{1}{18}(13x_2(0) + 8x_3(0)) - t$$

For the homogeneous case the general solution consists merely of the terms in (6.8) involving $x_2(0)$ and $x_3(0)$ with the others omitted. The solutions given here differ somewhat from those given by Campbell et al; this results from a trivial error made by them in the execution of their algorithm.

Of course this example is in some ways deceptively simple; however this is equally true of the solution obtained via the Drazin inverse. In general the system (6.7) above in which the matrix involving the derivatives is non-singular would be reached only after several stages of reduction (in fact k stages where k is the index associated with the relevant Drazin inverse). The solution of this reduced system can be expressed in terms of an exponential involving only an ordinary inverse.

REFERENCES

1 Ben-Israel, A and Greville, T. N. E., Generalized inverses-theory and application. New York, Wiley-Interscience, 1972.

2 Campbell, S. L., Meyer, C. D., and Rose, N. J., Applications of the Drazin inverse to linear systems of differential equations with singular constant coefficients. SIAM J. Appl. Math. 1976, 31, 411-425.

3 Van Dooren, P., The computation of Kronecker's canonical form of a singular pivot. To be published in Linear Algebra Appl.

4 Drazin, M. P., Pseudo inverses in associative rays and semigroups. Am. Math. Mon., 1958, 65, 506-514.

5 Gantmacher, F. R., The theory of matrices, Vol. II. New York, Chelsea, 1964.

6 Golub, G. H. and Wilkinson, J. H., Ill-conditioned eigensystems and the computation of the Jordan canonical form. SIAM Rev. 1976, 18, 578-619.

7 Greville, T. N. E., Spectral generalized inverses of square matrices. M R C Technical Science Rep. 823, Mathematics Research Center, University of Wisconsin, Madison, 1967.

8 Wilkinson, J. H., Linear Differential equations and Kronecker's canonical form. Invited paper at Symposium on Recent Advances in Numerical Analysis at Mathematics Research Center, Madison, 1978. To be published.

J. H. Wilkinson
Computer Science Department
Stanford University
Stanford, California 94305

K TANABE

Differential geometric approach to extended GRG methods with enforced feasibility in nonlinear programming: global analysis

ABSTRACT

A class of continuous analogues of 'extended GRG methods' for solving nonlinear programming problems is derived from a differential geometric point of view. In this connection, a continuous 'reduced Newton-Raphson method' for solving an underdetermined system of nonlinear equations is introduced. Then they are combined to obtain a class of continuous 'extended GRG methods' with special provision that corrects violation of constraint. Detailed analyses of the global behavior of these methods are given by using extensively the concept of generalized inverses. By applying a modified Newton-Raphson method and a quasi-Newton method to a particular setting of the problem, new constrained optimization algorithms are derived to enhance the convergence in a neighborhood of solutions of optimization problems.

1. INTRODUCTION

The methods for solving constrained optimization problems can be put into three categories:

1. Geometric methods such as Feasible Direction Method [54], Gradient Projection Method [42] and Generalized Reduced Gradient Method [1], in which successive approximations are chosen in such a way that the constraints remain satisfied. They are essentially ascent (or descent) methods with special provisions made against violation of constraints [18, 19, 21, 22, 33-36, 42, 44-49, 53-55].
2. Augmented function methods such as various forms of penalty function methods [11,13] and the Multiplier Method due to Hestenes [24] and Powell [38], in which, as opposed to the geometric methods that handle constraints rather directly, augmented performance indices with constant parameter are formed and then unconstrained optimization algorithms are successively applied to the indices for a sequence of values of parameters which are varied iteratively to improve the accuracy of approximations to the optimal solution [6, 7, 14, 15, 17,

18, 27, 32, 34, 39, 41, 43, 51]. They are closely associated with the methods in the following category.

3. Lagrangian Multiplier Methods in which active constraints at the optimal solution are identified somehow and are considered as equality constraints. Then nonlinear equation solvers such as Newton-Raphson method or Quasi-Newton methods [10,12] are applied to the system of nonlinear equations which are derived from the first order necessary conditions for the optimality [5, 10, 16, 18, 19, 25, 26, 28, 39, 51, 52].

Lagrangian Multiplier Methods are effective when a good initial estimate of the optimal solution is available since they generate rapidly convergent sequences of approximations to the solution. However, the convergence of these methods are guaranteed only for very restricted sets of starting points. Augmented Function Methods are designed to have larger regions of convergence. The multiplier methods [24,38,41] seem to be the most attractive among these methods, since they can avoid numerical difficulties that are frequently encountered with the usual penalty function methods. But their convergence region still depends on the choice of values of the penalty parameter in the augmented index, hence are liable to be affected with bad choice [34]. Besides, these methods have the disadvantage that they need to minimize the augmented functions on each iteration and their performance depends critically on those of the unconstrained optimization algorithms being employed, and it is not easy to ensure computationally the rapid convergence of unconstrained minimization algorithms for a large scale problem. In effect, augmented function methods are still local methods in philosophy, although they may have a fairly large region of convergence to the optimal solution for some problems.

Although less effort has been devoted to the development of the geometric methods recently, they have the convenient property that they converge to (locally) optimal solutions even from a very poor initial estimate of the solution and are useful at least for the initial stage of optimization processes. However, they have a serious disadvantage in that a correction procedure is frequently called for to restore the violation of constraints, which usually entails a considerable increase in computational time.

To alleviate the difficulties with the usual geometric methods, the present author [48,49] and Yamashita [54] developed a differential geometric method in which autonomous differential equations were introduced to define flows leading to optimal solutions, and obtained a class of algorithms by applying numerical integration processes to the autonomous system into which self-correcting features against violation of constraints were incorporated. The autonomous system can be interpreted as an amalgamation of continuous analogues of the gradient projection method and the Newton-Raphson method. The application of higher order methods such as the Runge-Kutta processes to the system allow for movement along arcs which are closer to the surface of the feasible region than the usual straight-line extrapolation methods, since they implicitly take into account the curvature of the surface to make extrapolation on the basis of more than one point. An important advantage of this approach is that we can avoid oscillation around the optimal solution which is often encountered with the usual straight-line methods when the contours of the objective function is closely tangential to the boundary of the constrained region around the solution.

Since the feasible set is a local attractor and an optimal point is an asymptotically stable point of the autonomous system if it satisfies the second order sufficient condition, we can use larger stepsizes for the numerical integration than usual without worrying about the violation of constraints. In fact, the Newton-Raphson correction term in the autonomous system can obviate the difficulty with the complicated and time-consuming treatment with violated constraints in a practical implementation of the usual method, and make it easy to control the stepsize automatically in the numerical integration process. An earlier method along this line is also found in the paper [45,46,47] which described a continuous analogue of the gradient projection method. An advantage of the differential geometric approach is that the global analysis of the method is facilitated, which gives us geometric insight into the global behavior of the discretized algorithms.

The purpose of this paper is (1) to extend the differential geometric method to the GRG method and obtain a class of 'extended GRG methods' with enforced feasibility, and (2) to develope a quasi-Newton algorithm and a modified Newton-Raphson algorithm for enhancing the convergence in the neighborhood of the optimal solution. The constitution of this paper parallels that of the previous paper [49]. In Section 2 the problem of a nonlinear constrained optimization problem is stated. In Section 3 a class of continuous analogues of 'extended GRG methods' is derived from the differential geometric point of view, and their global behaviors are analyzed. In Section 4 continuous analogue of 'reduced Newton-Raphson methods' for solving an underdetermined system of nonlinear equations are introduced and their global behavior is analyzed. In Section 5 the results of the previous two sections are combined to obtain the main method, which is the amalgamation of 'extended GRG methods' and 'reduced Newton-Raphson methods'. The asymptotic behavior of the basic autonomous system is fully described. In particular, the autonomous system is shown to be asymptotically stable at a regular optimal solution. In Section 6 by applying a quasi-Newton algorithm and a modified Newton-Raphson algorithm to the stationary condition of the autonomous system, new algorithms are derived to enhance the convergence in a neighborhood of a solution.

Throughout the paper we will use the terminology of the elementary theory of differentiable manifolds and dynamical systems [2,4,8,31]. We will also extensively use the concept of generalized inverses of a matrix and related projectors, which facilitates clear understanding of the geometric implications of formulas and equations.

2. STATEMENT OF PROBLEM

The problem considered in this paper is to maximize an objective function $f(x)$ with respect to $x \in R^n$ subject to the constraints,

$$g(x) \equiv (g_1(x), g_2(x), \dots, g_m(x))^t = 0 \in R^m, \qquad (1)$$

where R^n denotes the vector space formed by n-dimensional column vectors with real entries, $(*)^t$ denotes the transpose of $(*)$, and $f(x)$ and $g_i(x)$ are twice continuously differentiable. Although the problem contains only equality constraints, inequality constraints of the form

$$g_*(x) \leqq 0 \in R \qquad (2)$$

can be included by introducing new slack variables y_* such that

$$g_*(x) + y_*^2 = 0. \qquad (3)$$

This increases the number of critical points of the objective function with respect to the constraints. But created critical points can neither be maximal nor minimal points of the transformed problem. Maximal (minimal) points of the original problem exactly coincide with those of the transformed problem. One might suspect that the introduction of slack variables increases the computational work considerably. However by taking advantage of the simple structure of the equalities of the form (3) with respect to slack variables, we can reduce the computational time and storage required by the transformation much more than one might think, especially if we adopt the active constraints set strategy.

If we assume that the Jacobian $J_g(x) \equiv (\partial g_i/\partial x_j)$ is of full rank, i.e.,

$$\text{rank } J_g(x) = m, \tag{4}$$

on the feasible set V_g defined by

$$\{x \in R^n : g(x) = 0\} ,$$

then the feasible set V_g is an (n-m)-dimensional differentiable manifold in R^n and f is a differentiable function on the manifold V_g [2,4,31]. Figure 1 shows contours of the objective function on a manifold V_g schematically.

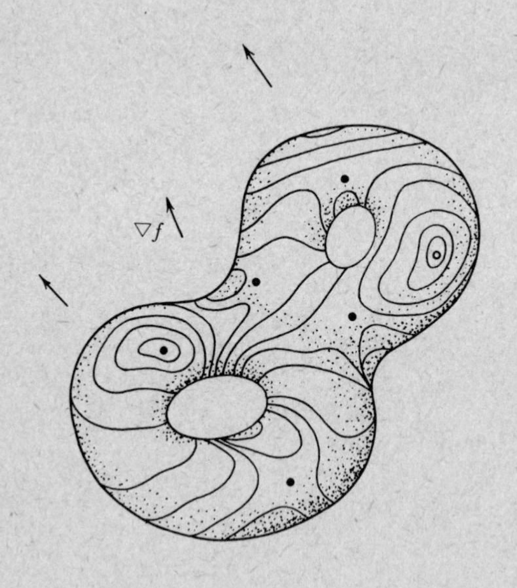

Fig. 1. Contours of f_v on V_g

The function f will be denoted by f_v when f is considered as a function on the manifold V_g and f_v will be distinguished from f. However we will not distinguish a point in V_g from the point in R^n. Thus our problem is to maximize f_v on the manifold V_g.

3. GENERAL CONTINUOUS REDUCED GRADIENT METHODS

In this section a differential geometric derivation is given to a class of continuous gradient projection methods which includes a continuous version of the GRG method as a special case. Global analysis of the behavior of the associated autonomous system is given. Different derivations of the GRG method are given, for example, in [1,33,44].

Throughout this section we assume that the condition (4) holds on V_g, i.e., V_g is a C^2 differentiable manifold. First and second derivatives of a function with respect to the variables

$$x \equiv (x_1, x_2, \ldots, x_n)^t \in R^n$$

will be represented by the operators

$$\nabla_x \equiv (\partial/\partial x_i) \quad \text{and} \quad \nabla_x^2 \equiv (\partial^2/\partial x_i \partial x_j)$$

respectively. The tangent space of the manifold V_g at a feasible point $x \in V_g$ will be denoted by T_x. T_x is represented by the null space

$$N(J_g(x)) \equiv \{z \in R^n : J_g(x) z = 0 \in R^m\} \tag{5}$$

of the Jacobian $J_g(x)$ of the mapping $g(x)$.

Let $\xi \equiv (\xi_1, \xi_2, \ldots, \xi_{n-m})^t$ be a local coordinate of a point $x \in V_g$ in a coordinate neighborhood $\tilde{U} \subset V_g$ of the point, i.e., let us assume that there exists a C^2-diffeomorphism γ between $\tilde{U} \subset V_g$ and an open set $\tilde{W} \subset R^{n-m}$, such that

$$\gamma : \tilde{U} \ni x \Longrightarrow \xi \equiv \gamma(x) \in \tilde{W},$$

and let us denote $\gamma^{-1}(\xi)$ by $x(\xi)$. Then

$$G(\xi) \equiv g(x(\xi)) = 0 \text{ for } \xi \in \tilde{W} \tag{6}$$

and

$$\operatorname{rank} \ J_x(\xi) = n - m, \tag{7}$$

where $J_x(\xi) \equiv (\partial x_i/\partial \xi_j)$ is the Jacobian matrix of the mapping $x(\xi)$. Differentiating both sides of (6) with respect to ξ, we have

$$J_G(\xi) = J_g(x)J_x(\xi) = 0 \text{ for } \xi \in \tilde{W}, \tag{8}$$

where $J_G(\xi)$ is the Jacobian matrix of $G(\xi)$. Letting

$$F(\xi) = f(x(\xi)), \tag{9}$$

and differentiating it with respect to ξ, we have

$$\nabla_\xi F(\xi) = J_x^t(\xi)\nabla_x f(x). \tag{10}$$

A continuous gradient method in terms of ξ is described by the autonomous system

$$\frac{d\xi}{dt} = \nabla_\xi F(\xi) = J_x^t(\xi)\nabla_x f(x). \tag{11}$$

By the chain rule, this sytem induces the autonomous system,

$$\begin{aligned}\frac{dx}{dt} &= J_x(\xi)\ \frac{d\xi}{dt} \\ &= J_x(\xi)J_x^t(\xi)\nabla_x f(x),\end{aligned} \tag{13}$$

in terms of x, where $J_x(\xi)$ is subject to the conditions (7) and (8). It should be noted here that

$$\begin{aligned}\frac{dg(x)}{dt} &= J_g(x)\ \frac{dx}{dt} \\ &= (J_g(x)J_x(\xi))J_x^t(\xi)\nabla_x f(x) = 0\end{aligned} \tag{14}$$

by virtue of (8).

Since $J_g(x)$ is of full rank on V_g, for any point $x \in V_g$, there exists a neighborhood $U \subset R^n$ of the point and an $n \times n$ permutation matrix π such that

$$J_g(x)\pi \equiv [A(x) \mid B(x)] \tag{15}$$

and the $m \times m$ matrix $A(x)$ is nonsingular on U. We always assume that U is the largest set among those which satisfies this condition. Letting

$$\pi^t J_x(\xi) = \begin{bmatrix} C(x) \\ --- \\ D(x) \end{bmatrix} \tag{16}$$

where $D(x)$ is an $(n-m) \times (n-m)$ matrix, we have by (8),

$$J_g(x)J_x(\xi) = A(x)C(x) + B(x)D(x) = 0, \tag{17}$$

from which we have

$$C(x) = -A^{-1}(x)B(x)D(x). \tag{18}$$

Hence

$$J_x(\xi) = \pi \begin{bmatrix} -A^{-1}(x)B(x) \\ \hline I_{n-m} \end{bmatrix} D(x) \tag{19}$$

where $D(x)$ is an $(n-m) \times (n-m)$ nonsingular matrix by (7). Thus

$$J_x(\xi)J_x^t(\xi) = P_G(x)D_G(x)D_G^t(x)P_G^t(x) \tag{20}$$

where $P_G(x)$ is an $n \times n$ projection matrix onto $N(J_g(x))$, defined by

$$P_G(x) \equiv \pi \left[\begin{array}{c|c} 0 & -A^{-1}(x)B(x) \\ \hline 0 & I_{n-m} \end{array}\right] \pi^t, \tag{21}$$

and $D_G(x)$ is an $n \times (n-m)$ matrix of rank $n-m$, defined by

$$D_G(x) \equiv \pi \begin{bmatrix} 0 \\ \hline D(x) \end{bmatrix} . \tag{22}$$

If we define an n m matrix $\tilde{J}_g^-(x)$ by

$$\tilde{J}_g^-(x) \equiv \pi \begin{bmatrix} A^{-1}(x) \\ \hline 0 \end{bmatrix} , \tag{23}$$

then $\tilde{J}_g^-(x)$ is a reflexive generalized inverse of $J_g(x)$, i.e.,

$$J_g(x)\tilde{J}_g^-(x)J_g(x) = J_g(x) \quad \text{and} \quad \tilde{J}_g^-(x)J_g(x)\tilde{J}_g^-(x) = \tilde{J}_g^-(x)$$

$P_G(x)$ is represented as

$$P_G(x) = I_n - \tilde{J}_g^-(x)J_g(x), \tag{24}$$

and

$$J_x(\xi) = P_G(x)D_G(x). \tag{25}$$

Substituting (20) into (13), we have the basic autonomous system

$$\frac{dx}{dt} = \Phi_G(x) \equiv J_x(\xi)J_x^t(\xi)\nabla_x f(x)$$
$$= P_G(x)D_G(x)D_G^t(x)P_G^t(x)\nabla_x f(x), \tag{26}$$

which will be called a 'general continuous gradient acute projection method', since

$$\langle \nabla_x f(x), \Phi_G(x) \rangle = ||J_x^t(\xi)\nabla_x f(x)||^2 \geqq 0. \tag{27}$$

Associated with $\Phi_G(x)$, there exists a differentiable mapping $\Lambda_G(x)$ from $U \cap V_g$ to R^m such that

$$\left[\begin{array}{c|c} M(x) & J_g^t(x) \\ \hline J_g(x) & 0 \end{array}\right] \left[\begin{array}{c} \Phi_G(x) \\ \hline \Lambda_G(x) \end{array}\right] = \left[\begin{array}{c} \nabla_x f(x) \\ \hline 0 \end{array}\right], \tag{28}$$

where M(x) is the n×n nonnegative definite matrix defined by

$$M(x) \equiv (D_G(x)D_G^t(x))^\dagger = \pi \left[\begin{array}{c|c} 0 & 0 \\ \hline 0 & (D(x)D^t(x))^{-1} \end{array}\right] \pi^t.$$

By (28), we have

$$\Lambda_G(x) = (\tilde{J}_g^-(x))^t \nabla_x f(x). \tag{29}$$

It should be noted that rank M(x) = n−m.

Different choices of π and D(x) (or M(x)) in (26) produce different methods. A continuous version of the GRG method is a special case of (26) (and (13)), where $D(x) = I_{n-m}$. To see this, let us assume that x admits a partition of elements,

$$x = \begin{bmatrix} x_B \\ x_I \end{bmatrix} \begin{matrix} m \\ n-m \end{matrix},$$

such that the induced partition,

$$J_g(x) \equiv [(J_g(x))_B \mid (J_g(x))_I] \tag{30}$$

satisfies the condition that $(J_g(x))_B$ is nonsingular in some neighborhood U. Then $x_I \in R^{n-m}$ forms a local coordinate of V_g in $V_g \cap U$. The vectors $x_B \in R^m$ and $x_I \in R^{n-m}$ are called 'basic' variables and 'independent' variables respectively. In this case, we have

$$J_G(x_I) = J_g(x)J_x(x_I) = J_g(x) \begin{bmatrix} (J_x(x_I))_B \\ \hline (J_x(x_I))_I \end{bmatrix}$$

$$= J_g(x) \begin{bmatrix} (J_x(x_I))_B \\ \hline I_{n-m} \end{bmatrix} = (J_g(x))_B(J_x(x_I))_B + (J_g(x))_I$$

$$= 0 \tag{31}$$

which corresponds to (17). Hence

$$J_x(x_I) = \begin{bmatrix} -(J_g(x))_B^{-1}(J_g(x))_I \\ \hline I_{n-m} \end{bmatrix} . \tag{32}$$

Substituting (32) into (26), we obtain a continuous version of GRG method

$$\frac{dx}{dt} = \Phi(x) \equiv \begin{bmatrix} -(J_g(x))_B^{-1}(J_g(x))_I \\ \hline I_{n-m} \end{bmatrix} \begin{bmatrix} -(J_g(x))_B^{-1}(J_g(x))_I \\ \hline I_{n-m} \end{bmatrix}^t \nabla_x f(x)$$

$$= P(x)P^t(x)\nabla_x f(x) \tag{33}$$

where the n×n matrix P(x) is a projector onto $N(J_g(x))$, defined by

$$P(x) \equiv \left[\begin{array}{c:c} 0 & -(J_g(x))_B^{-1}(J_g(x))_I \\ \hline 0 & I_{n-m} \end{array}\right] . \tag{34}$$

If we define an n×m matrix $J_g^-(x)$ by

$$J_g^-(x) \equiv \begin{bmatrix} (J_g(x))_B^{-1} \\ \hline 0 \end{bmatrix} , \tag{35}$$

then $J_g^-(x)$ is a reflexive generalized inverse of $J_g(x)$ and P(x) is represented as

$$P(x) = I_n - J_g^-(x) J_g(x). \tag{36}$$

Associated with Φ(x), there exists a differentiable mapping Λ(x) from $V_g \cap U$ to R^m such that

$$\left[\begin{array}{cc|c} 0 & 0 & \\ \hline 0 & I_{n-m} & J_g^t(x) \\ \hline \multicolumn{2}{c|}{J_g(x)} & 0 \end{array}\right] \begin{bmatrix} \Phi(x) \\ \\ \hline \Lambda(x) \end{bmatrix} = \begin{bmatrix} \nabla_x f(x) \\ \\ \hline 0 \end{bmatrix} \tag{37}$$

from which we have

$$\Lambda(x) = ((J_g(x))_B^{-1})^t \nabla_{x_B} f(x)$$

$$= (J_g^-(x))^t \nabla_x f(x) , \tag{38}$$

where $J_g^-(x)$ is the reflexive generalized inverse (35).

It should be noted here that (33) is a special case of (26), where $\pi \equiv I_n$, $D \equiv I_{n-m}$. A more desirable choice of D_G will be described in Section 5.

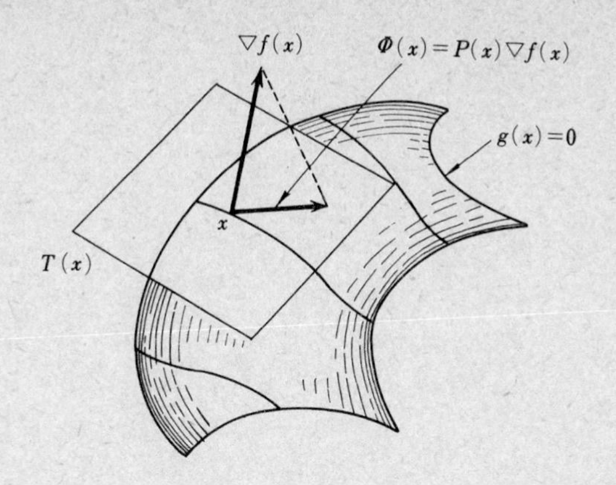

Figure 2. Gradient Acute Projection

It is easily seen from the forms of (10) and (13) that the set $\{x \in V_g : \Phi_G(x) = 0 \in R^n\}$ of equilibrium points of the autonomous system (13) (and (26)) and coincides with the set

$$C_f \equiv \{x \in V_g : (df_v)_x = 0\}$$

of critical points of the function f_v, where $(df_v)_x$ is the differential of f_v at the point x. A critical point $x \in C_f$ of f_v is called regular if the Hessian $(\partial^2 f_v/\partial\xi_i \partial\xi_j)$ of f_v is nonsingular at x, where $\{\xi_i\}$ (i = 1,2,...,n-m) is a local coordinate system at $x \in V_g$.

Theorem 3.1 If $J_x(\xi)$ is continuously differentiable in ξ, i.e., if D(x) in (19) is continuously differentiable, then for any feasible point $x^o \in V_g \cap U$, there exists a unique solution of (13) (and (26)) defined in a interval $0 \leq t < M$ with the initial condition, $x(0,x^o) = x^o$, such that

1. the trajectory $x(t,x^o)$, $0 \leq t < M$ remains in the feasible manifold V_g, i.e., V_g is an invariant set of the autonomous system,
2. if x^o is not a critical point, then for $0 \leq t < M$,

$$\frac{df(x(t,x^o))}{dt} > 0 \tag{39}$$

Proof The existence of a solution is easily seen from the elementary theory of differential equations. We have from (14) that

$$\frac{dg(x(t,x^o))}{dt} = 0$$

hence, $g(x(t,x^o)) \equiv g(x^o) = 0$. Finally, we have

$$\begin{aligned}\frac{df(x(t,x^o))}{dt} &= \langle \nabla_x f(x), dx/dt \rangle \\ &= \langle \nabla_x f(x), J_x(\xi) J_x^t(\xi) \nabla f(x) \rangle \\ &= ||J_x^t(\xi)\nabla_x f(x)||^2 \\ &= ||D_G^t(x) P_G^t(x) \nabla_x f(x)||^2 \geqq 0. \end{aligned} \qquad (40)$$

Suppose $||D_g^t(x)P_G^t(x)\nabla f(x)||$ vanishes at $\tilde{t} \in [0\ M)$, then

$$\Phi_G(x(\tilde{t},x^o)) = 0 \in R^n.$$

Hence $x(\tilde{t},x^o)$ is an equilibrium point of the system (13) (or (26)). On the other hand $x(t) \equiv x(\tilde{t},x^o)$, $0 \leqq t < M$, is obviously a solution of the system (26). Thus by the uniqueness of the solution of (26), we have
$x(t,x^o) \equiv x(\tilde{t},x^o) = x^o$; which implies x^o is an equilibrium point of (26).□

In this paper a solution $x(t,x)$, $0 \leqq t < M$ of an autonomous system means the one which cannot be extended to a larger interval in positive direction of 'time' t for a given initial point x^o.

The proofs of the following theorems are very similar to those given in [46,47].

Theorem 3.2 Let $x^o \in U \cap V_g$ be a noncritical point. If the solution $x(t,x^o)$, $0 \leqq t < M$ of (26) is bounded and if it does not approach the boundary ∂U of U, then $M = \infty$ and the positive limit set Γ is a compact connected set contained in the boundary ∂C_f of critical set C_f.

Proof Since the solution lies in the manifold V_g, it follows from the assumption that the positive limit set Γ of the solution is nonempty, invariant and compact in the manifold V_g, hence $M = \infty$. Let K be a compact set in the manifold V_g which contains the trajectory $x(t,x^o)$, $t \in [0,\infty)$. From the continuity of f on V_g, it follows that f is bounded on K. As is seen in Theorem 3.1, $f(x(t,x^o))$ is a strictly monotone increasing function of $t \in [0,\infty)$. It follows that $f(x(t,x^o))$ converges to $\tilde{\ell} < \infty$ as $t \to \infty$. Again from the continuity of f, it follows that $f(x) = \tilde{\ell}$ for $x \in \Gamma$.

Let $\tilde{x}$ be a point in the positive limit set Γ. Since $\Gamma \subset V_g$ is an invariant set of the system (26), there exists a solution $x(t,\tilde{x})$ defined in some interval $[0,N)$ with the initial point $\tilde{x}$ such that $x(t,\tilde{x}) \in \Gamma$ for $t \in [0,N)$.

If x is not a critical point, then it follows from Theorem 3.1 that $f(x(t,x))$ is a strictly monotone increasing function of $t \in [0,N)$. But this contradicts the fact that f(x) is constant on Γ. Thus Γ is contained in C_f. It is easily seen that Γ is connected (See [8]). It is easily seen that $\Gamma \subset \partial C_f$, so the proof is omitted. □

It follows from the theorem that if the solution converges to a point $\tilde{x}$, then $\tilde{x}$ is a critical point of the function f on the manifold V_g.

<u>Corollary</u> If $C_f \cap K$ consists of isolated critical points of f, besides the assumption of Theorem 3.2 being satisfied, then the solution $x(t,x^o)$ converges to a critical point of f as t tends to infinity, i.e.

$$\lim_{t\to\infty} x(t,x^o) = \tilde{x} \in C_f.$$

The result is easy to establish, so the proof is omitted. It should be noted here that a regular critical point is an isolated critical point.

<u>Theorem 3.3</u> Let $\hat{C} \subset U$ be a compact connected component of C_f such that

$$\hat{C} \cap \overline{(C_f - \hat{C})} = \phi ,$$

where $\overline{C}$ denotes the topological closure of C. If the function f is maximal on the set $\hat{C}$, then $\hat{C}$ is asymptotically stable set of the system (26).

<u>Proof</u> Let $f(x) = \mu$ on $\hat{C}$. There exists a relatively compact neighborhood $U_{\hat{C}} \subset V_g$ of $\hat{C}$ such that

$$U_{\hat{C}} \cap (C_f - \hat{C}) = \phi,$$

and

$$\mu > f(x) \text{ for } x \in \overline{U}_{\hat{C}} - \hat{C} .$$

Let μ' be a maximal value of f(x) on the boundary of $U_{\hat{C}}$. Then $\mu > \mu'$. Since f(x) is continuous on V_g, we can choose a neighborhood $V_{\hat{C}} \subset U_{\hat{C}}$ of $\hat{C}$ such that

$$f(x) > (\mu+\mu')/2 \text{ for } x \in V_{\hat{C}} .$$

Then for any solution $x(t,x^o)$ of (26) defined in $[0,M)$ with the initial point $x^o \in V_{\hat{C}}$, we have

$$f(x(t,x^o)) > (\mu+\mu')/2 \text{ for } t \in [0,M).$$

Thus the solution of (26) which starts from a point in the neighborhood $V_{\hat{C}}$ remains in the relatively compact set $U_{\hat{C}}$. Next from Theorem 3.2, it follows that the solution approaches a critical set in $\overline{U}_{\hat{C}}$ as $t \to \infty$. On the other hand,

$$C_f \cap \overline{U}_{\hat{C}} = \hat{C}$$

so that the solution approaches $\hat{C}$. □

The next corollary follows immediately from the theorem.

Corollary If an isolated critical point p is a maximal point of f on V_g, then the system (26) is asymptotically stable at the point p in the manifold V_g.

Note that if a solution hits the boundary ∂U of U in V_g, then new basic variables should be chosen so that the process is continued.

4. CONTINUOUS REDUCED NEWTON-RAPHSON METHOD

We consider the problem of solving the system

$$g(x) \equiv (g_1(x), g_2(x), \ldots, g_m(x))^t = 0 \in R^m \tag{40}$$

of nonlinear equations. Our problem is to find a feasible point.

Branin [9] considered the autonomous system

$$dg(x)/dt = -g(x), \tag{41}$$

or equivalently

$$J_g(x)\ (dx/dt) = -g(x), \tag{42}$$

for solving the system (40) in the case where $m = n$. The present author extended this method to the case where $m \leq n$ and analyzed its global behavior [49,50]. In this section another extension of the method is given in preparation for the discussion of Section 5 in connection with the continuous version of the GRG method.

Given an initial point x^o, if a vector-valued function $x(t,x^o)$ defined in an interval $0 \leq t < M$, such that $x(0,x^o) = x^o$, satisfies the equation (41), it will be called a solution of the autonomous system (41). It is easily seen that a solution of the system satisfies a 'first integral'

$$g(x(t,x^o)) = \exp(-t)g(x^o) \text{ for } 0 \leq t < M. \tag{43}$$

Hence the image of the trajectory of a solution of the system by the mapping g lies on the line segment between the point $g(x^o)$ and the origin of R^m. More precisely, the point $g(x(t,x^o))$ moves along the straight line toward the origin as the 'time' proceeds. If a solution exists for the interval $0 \leq t < \infty$

$$\lim_{t\to\infty} g(x(t,x^o)) = 0 \in R^m. \tag{44}$$

It is natural to expect that the solution approach the set,

$$V_g \equiv \{x \in R^n : g(x) = 0 \in R^m\} \tag{45}$$

of solutions of (40). We will establish under what condition this holds.

In the open set U defined in (15), the system (42) is equivalent to the system

$$\frac{dx}{dt} = -\tilde{J}_g^-(x)g(x) + (I_n - \tilde{J}_g^-(x)J_g(x))h(x) \tag{46}$$

so long as the condition

$$\text{rank } J_g(x) = m \tag{47}$$

is satisfied, where $\tilde{J}_g^-(x)$ is the reflexive generalized inverse of $J_g(x)$ given in (23) and $h(x)$ is an arbitrary mapping from R^n to R^n. In particular, if x admits a partition of variables x such as (30), then the system (42) is also equivalent to

$$\frac{dx}{dt} = -J_g^-(x)g(x) + (I_n - J_g^-(x)J_g(x))h(x) \tag{48}$$

where $J_g^-(x)$ is the reflexive generalized inverse of $J_g(x)$ given in (35).

<u>Lemma 4.1</u> If rank $J_g(x^o) = m$ for some $x^o \in U$ and $h(x)$ is sufficiently smooth, then there exists a solution $x(t,x^o)$ of the system (46) (or(48)) for some interval $0 \leq t < M$, which can be prolonged until its trajectory approaches either the set,

$$S \equiv \{x \in R^n : \text{rank } J_g(x) < m\} \ ,$$

of singular points or the boundary ∂U of U; otherwise $M = \infty$.

Proof By the assumption, $J_g(x)$ is full rank in a neighborhood $U' \subset U$ of x^o. Hence the system (42) and (46) (or (48)) are equivalent in U'. Noting that the righthand side of the system (46) (or (48)) is continuous in U' and that U' may be the connected component of U separated by the set S, which contains x^o, we obtain the desired result by the elementary theory of differential equations. □

Now consider the case where $h(x) \equiv 0$ in (46), i.e.

$$\frac{dx}{dt} = -\tilde{J}_g^-(x)g(x) = -\pi \begin{bmatrix} A^{-1}(x)g(x) \\ \hline 0 \end{bmatrix} . \tag{49}$$

Lemma 4.2 If rank $J_g(x^*) = m$ for a feasible point $x^* \in U \cap V_g$, then x^* is a stable equilibrium point of the system (49).

Proof Without loss of generality, we assume that $\pi = I$. Then (49) is equivalent to

$$\frac{dx_B}{dt} = \Phi_B(x) = -(J_g(x))_B^{-1} g(x), \tag{50}$$

and

$$\frac{dx_I}{dt} = 0, \text{ where } \quad x = \begin{bmatrix} x_B \\ \cdots \\ x_I \end{bmatrix} \begin{matrix} m \\ \\ n-m. \end{matrix}$$

Since $x_I \equiv x_I^0$, by (50), where

$$x^0 = \begin{bmatrix} x_B^0 \\ \hline x_I^0 \end{bmatrix} , \quad \text{we have}$$

$$\Phi_B(x) = -(J_g(\begin{bmatrix} x_B \\ \hline x_I^0 \end{bmatrix}))_B^{-1} g(\begin{bmatrix} x_B \\ \hline x_I^0 \end{bmatrix}) \quad . \tag{51}$$

Hence the value of Jacobian matrix $J_{\Phi_B}(x)$ of $\Phi_B(x)$ with respect to the variables x_B at the point,

$$x^* = \begin{bmatrix} x_B^* \\ x_I^* \end{bmatrix}, \text{ is given by}$$

$$J_{\Phi_B}(x^*) = -(J_g(x^*))_B^{-1}(J_g(x^*))_B = -I_m \quad . \tag{52}$$

Thus by Liapunov's direct method, the system,

$$\frac{dx}{dt} = \Phi_B \begin{bmatrix} x_B \\ \hline 0 \\ x_I \end{bmatrix} \tag{53}$$

is asymptotically stable at the point x_B^* . □

Theorem 4.3 If rank $J_g(x^*) = m$ for a feasible point $x^* \in U \cap V_g$, then there exists a neighborhood U^* of x^* such that the solution $x(t,x^o)$ of (49) which starts from any point x^o in U^* exists for the interval $0 \leq t < \infty$, and always converges to a feasible point, which is not necessarily x^*, as t tends to infinity.

Proof By the previous lemma, there exists a neighborhood $U^* \subset U$ such that the solution $x(t, x^o)$ which starts from a point x^o in U^* is bounded in a compact set $W \supset U^*$ where $W \subset U$. Hence it exists for $0 \leq t < \infty$, by lemma 4.1. Since U^* can be chosen so that rank $J_g(x) = m$ for $x \in W$, $J_g^-(x)$ is continuous in the compact set W. Therefore there exists a positive number k such that

$$||J_g^-(x)|| < k \text{ for } x \in W.$$

Since the solution $x(t, x^o)$ satisfies the equality, (43)

$$g(x(t, x^o)) = \exp(-t)g(x^o) \text{ for } 0 \leq t < \infty$$

we have

$$\int_0^\infty ||dx/dt||dt = \int_0^\infty ||J_g^-(x(t,x^o))g(x(t,x^o))||dt$$

$$< k \int_0^\infty ||g(x(t,x^o))||dt$$

$$= k||g(x^o)||\int_0^\infty \exp(-t)dt = k||g(x^o)|| \; .$$

This implies that the length of the trajectory is finite, thus $x(t, x^o)$ must converge to a point $\bar{x}$ which is feasible point by (43).

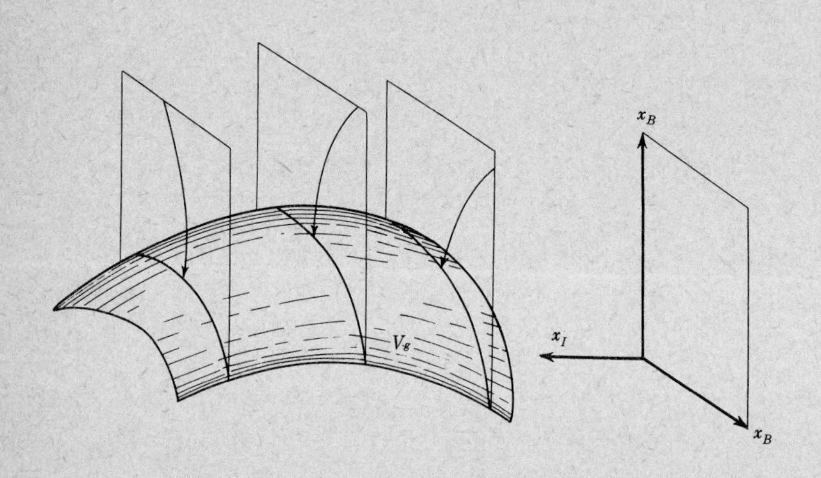

Figure 3. Trajectories of the System (49)

The following theorem describes the global behavior of the system (49).

Theorem 4.4 For a given $x^o \in S$ there exists a solution $x(t, x^o)$, $0 \leq t < M$, of (49) with $x(t, x^o) = x^o$. As t tends to M, its trajectory will either approach the set $S \cup \partial U$ or converge to a feasible point $x^* \in V_g \cap S^c$, in which case $M = \infty$ and

$$||x(t,x^o) - x^*|| = k|| g(x^o)|| \quad \exp(-t), \ 0 \leq t < \infty .$$

Proof If the trajectory does not approach $S \cup \partial U$, then by Lemma 4.1, we have $M = \infty$. Let Γ be the positive limit set of the solution. Then by the relation (43), Γ is contained in the feasible set V_g. Hence, by Theorem 4.3, the trajectory is contained in a compact set W such that

$W \cap S = \emptyset$. Since $J_g^-(x)$ is continuous mapping, then there exists a constant k such that

$$||J_g^-(x)|| < k \text{ for } x \in W.$$

Thus we have

$$||x(t,x^o) - x^*|| = ||\int_t^\infty (dx/dt)dt|| \leqq \int_t^\infty ||dx/dt|dt$$

$$= \int_t^\infty ||J_g^-(x)g(x)||dt \leqq k\int_t^\infty ||f(x(t,x^o))||dt$$

$$= k||f(x^o)||\int_t^\infty \exp(-t)\ dt = k||g(x^o)||\ \exp(-t). \quad \square$$

If a solution approachs the boundary ∂U of U, then a different partition of the variable x should be taken to continue the process for obtaining a feasible point.

5. CONTINUOUS REDUCED GRADIENT METHODS WITH ENFORCED FEASIBILITY

The continuous gradient acute projection methods need a correction procedure for restoration of violation of constraints. Fortunately however, a self-correcting feature can be incorporated into the basic systems (26) and (33) by combining the results in the previous two sections. An earlier version of one of these methods is suggested by using a different approach in [22].

The continuous gradient acture projection method is described by

$$\frac{dx}{dt} = \Phi_G(x),$$

where

$$M(x)\Phi_G(x) + J_g^t(x)\Lambda_G(t) = \nabla_x f(x) \tag{54}$$

and

$$J_g(x)\ \Phi_G(x) = 0\ . \tag{55}$$

The condition (55) implies

$$\frac{dg(x)}{dt} = J_g(x)\ \frac{dx}{dt} = J_g(x)\Phi_G(x) = 0, \tag{56}$$

which also means that $\Phi_G(x)$ belongs to the tangent space T_x of the manifold at the points x. If we impose the condition

$$\frac{dg(x)}{dt} = -g(x) \tag{57}$$

instead of (55), i.e., if we determine $\Phi_G(x)$ by

$$\left[\begin{array}{c|c} M(x) & J_g^t(x) \\ \hline J_g(x) & 0 \end{array}\right] \left[\begin{array}{c} \Psi_G(x) \\ \hline \bar{\Lambda}_G(x) \end{array}\right] = \left[\begin{array}{c} \nabla_x f(x) \\ \hline -g(x) \end{array}\right], \tag{58}$$

then we obtain the autonomous system,

$$\begin{aligned} \frac{dx}{dt} = \Psi_G(x) &\equiv -\tilde{J}_g(x)g(x) + J_x(\xi)\ J_x^t(\xi)\nabla_x f(x) \\ &= -\tilde{J}_g^-(x)g(x) = (I-\tilde{J}_g^-(x)J_g(x))D_G(x)D_G^t(x)P_G^t(x)\nabla_x f(x) \end{aligned} \tag{59}$$

where $J_x(\xi)$, $\tilde{J}_g^-(x)$ and $P_G(x)$ are given in (25), (23) and (21) respectively. It should be noted that $\Lambda_G(x)$ has identical expression,

$$\bar{\Lambda}_G(x) = (\tilde{J}_g^-(x))^t\nabla_x f(x)$$

to (29), but it is defined in $U \subset R^n$. The basic system (59) is the continuous gradient acute projection method (25) with an additional correction term $-\tilde{J}_g^-(x)g(x)$.

Corresponding to the partition (30) of x, we obtain a continuous reduced gradient method with a correction term.

$$\begin{aligned} \frac{dx}{dt} = \Psi(x) &\equiv -J_g^-(x)g(x) + P(x)P^t(x)\nabla_x f(x) \\ &= -J_g^-(x)g(x) + (I-J_g^-(x)J_g(x))(I-J_g^-(x)J_g(x))^t\nabla_x f(x) \end{aligned} \tag{60}$$

which is a special case of (59), where $J_g^-(x)$ is given in (35). Associated with (60), there exists a mapping $\bar{\Lambda}(x)$ defined from U to R^m such that

$$\left[\begin{array}{cc|c} 0 & 0 & \\ \hline 0 & I_{n-m} & J_g^t(x) \\ \hline \multicolumn{2}{c|}{J_g(x)} & 0 \end{array}\right] \left[\begin{array}{c} \Psi(x) \\ \hline \bar{\Lambda}(x) \end{array}\right] = \left[\begin{array}{c} \nabla f(x) \\ \hline -g(x) \end{array}\right]. \tag{61}$$

The mapping $\Psi_G(x)$ (or $(\Psi(x))$ and $\Phi_G(x)$ (or $\Phi(x)$) are identical on V_g. Hence the solutions of (59) (or (60)) and (26) (or (33)) are identical if they have the same initial points in V_g. The basic system (59) is defined in U except for the set S of singular points. In fact, the system (59) (or (60)) is a special case of the system (46) (or (48)) where

$$h(x) \equiv D_g(x)D_G^t(x)(I-\tilde{J}_g^-(x)J_g(x))^t\nabla_x f(x), \tag{62}$$

$$(\text{or } h(x) \equiv (I-J_g^-(x)J_g(x))^t\nabla_x f(x)). \tag{63}$$

Hence Lemma 4.1 applies to the systems (59) and (60) and their solution satisfies the equation (43). In other words, their trajectories will either approach the set $S \cup \partial U$, or $M = \infty$. And their trajectories are pulled towards V_g as time t proceeds.

Theorem 5.1 If a solution $x(t,x^o)$, $0 \leq t < M$ of the system (59) (or (60)) is bounded and does not approach the set $S \cup \partial U$, then $M = \infty$ and the positive limit set Γ is a compact connected set contained in the set C_f of initial points of f_v.

Proof By Lemma 4.1 we have $M = \infty$. Since the solution is bounded and satisfies the equation (43), its positive limit set Γ is a compact connected set contained in V_g. We have

$$\begin{aligned} df(x(t,x^o))/dt &= \langle \nabla f(x), dx/dt \rangle \\ &= \langle \nabla f(x), -\tilde{J}_g^-(x)g(x) \rangle + \langle \nabla f(x), J_x(\xi)J_x^t(\xi)\nabla f(x)\rangle \\ &= \langle \nabla f(x), -\tilde{J}_g^-(x)g(x) \rangle \\ &= \exp(-t)\langle \nabla f(x(t,x^o)), -J_g^-(x(t,x^o))g(x^o) \rangle, \end{aligned}$$

since $\langle \nabla f(x), J_x(\xi)J_x^t(\xi)\nabla f(x)\rangle = ||J_x^t(\xi)\nabla f(x)||^2 \geq 0$,

and $x(t,x^o)$ satisfies the equation (43), where $J_x(\xi)$ and $P_G(x)$ are defined in (19) and (21), respectively. Under the assumption, there exists a compact set W which contains the trajectory of the solution and does not intersect with the set $S \cup \partial U$. Since $\langle \nabla f(x), -J_g^-(x)g(x^o) \rangle$ is continuous on W, it is uniformly bounded on W, i.e., $|\langle \nabla f(x), -J_g^-(x)g(x) \rangle | < \nu$ for $x \in W$, for some positive number k. Hence

$$\frac{df(x(t,x^o))}{dt} = -\nu \exp(-t). \tag{64}$$

Suppose there exist two points x_1 and x_2 in the positive limit set Γ such that $f(x_1) > f(x_2)$. Then there exist neighborhood U_1 and U_2 of x_1 and x_2 respectively such that $f(x) > f(x_1) - \delta$ for $x \in U_1$, and $f(x) < f(x_2) + \delta$ for $x \in U_2$, where $\delta \equiv (f(x_1) - f(x_2))/3$. Since x_1 and x_2 belong to Γ, there exist two positive numbers t_1 and $t_2 (t_1 < t_2)$ such that

$$\nu\exp(-t_1) < \delta,\ x(t_1,x^o) \in U_1 \text{ and } x(t_2,x^o) \in U_2.$$

Intergrating both sides of the inequality (64), we have

$$f(x(t_2,x^o)) - f(x(t_1,x^o)) = \int_{t_1}^{t_2} df(x(t,x^o))$$

$$\geqq \int_{t_1}^{t_2} -\nu \exp(-t)dt$$

$$= \nu \exp(-t_2) - \nu \exp(-t_1)$$

$$> -\nu \exp(-t_1) > -\delta. \tag{65}$$

But we have $f(x(t_1,x^o)) > f(x_1) - \delta$ and $f(x(t_2,x^o)) < f(x_2) + \delta$, which contradict (65). Thus the function f is constant on the positive limit set Γ. Let $\bar{x}$ be a point in the set Γ. Since Γ is an invariant set of the system (59), there exists a solution $x(t,\bar{x})$, $0 \leqq t < M'$ with the initial point $\bar{x}$, such that $x(t,\bar{x}) \in \Gamma$ for $0 \leqq t < M'$. Suppose $\bar{x}$ is not a critical point of f_v. Since $x(t,\bar{x})$ is a solution of the system (26), it follows from Theorem 3.1 that $f(x(t,\bar{x}))$ is a strictly monotone increasing function of t. But this contradicts the fact that $f(x)$ is constant on Γ. Thus Γ is contained in C_f. □

The following theorem will partly guarantee the stability of our algorithms in a neighborhood of the solution.

<u>Theorem 5.2</u> The system (59) is asymptotically stable at a regular maximal point $\bar{x} \in C_f \cap U$ of the function f_v, if rank $J_g(\bar{x}) = m$.

Before proving the theorem we prepare a few lemmas. Let a block matrix $[AB_1C: AB_2C: AB_3C: \dots : AB_kC]$ be denoted by

$$A[B_1:B_2:B_3: \dots :B_k]C,$$

where A, B_i's and C are suitable vectors or matrices, and let a block matrix

$[A_1^t:A_2^t:A_3^t: \ldots :A_k^t]$ be denoted by $[A_1:A_2:A_3: \ldots :A_k]^t$.

Lemma 5.3 The Jacobian $J_{\Psi_G}(x)$ of the mapping $\Psi_G(x)$ defined in (59) is given by

$$
\begin{aligned}
J_{\Phi_G}(x) &= -\tilde{J}_g^-(x)J_g(x) \\
&+ P_G(x)D_G(x)D_G^t(x)P_G^t(x)(H_f(x) - \sum_{i=1}^m \lambda_i(x)H_{g_i}(x)) \\
&- S_g(x)g(x) + T_g(x)J_x^t(\xi)\nabla_x f(x) \\
&+ J_x(\xi)Z_g^t(x)P_G^t(x)\nabla_x f(x),
\end{aligned} \tag{66}
$$

as long as rank $J_g(x) = m$, where $H_f(x)$ and $H_{g_i}(x)$ are Hessians of f and g_i respectively, i.e., $H_f(x) \equiv \nabla^2 f(x)$ and $H_{g_i}(x) \equiv \nabla^2 g_i(x)$,

$$\Lambda(x) \equiv (\lambda_i(x)) = (J_g^-(x))^t \nabla f(x),$$

$$R_g(x) \equiv [\partial J_g(x)/\partial x_1: \ldots : \partial J_g(x)/\partial x_n],$$

$$S_g(x) \equiv [\partial \tilde{J}_g^-(x)/\partial x_1: \ldots : \partial J_g^-(x)/\partial x_n],$$

$$T_g(x) \equiv [\partial J_x(\xi)/\partial x_1: \ldots : \partial J_x(\xi)/\partial x_n],$$

$$Z_g(x) \equiv [\partial D_G(x)/\partial x_1: \ldots : \partial D_G(x)/\partial x_n],$$

$P_G(x)$, $\tilde{J}_g^-(x)$, $J_x(\xi)$ and $D_G(x)$ are defined in (21), (23), (19) and (22) respectively.

Proof We will drop the arguments x and ξ in proofs only. Differentiating the equality,

$$\Psi_G = -\tilde{J}_g^- g + J_x J_x^t \nabla f,$$

with respect to x, we have

$$J_{\Psi_G} = -S_g g - \tilde{J}_g^- J_g + T_g J_x^t \nabla f + J_x T_g^t \nabla f + J_x J_x^t H_f. \tag{67}$$

Since

$$\partial \tilde{J}_g^- / \partial x_j = \pi \begin{bmatrix} \partial A^{-1}/\partial x_j \\ \hline 0 \end{bmatrix}$$

$$= \pi \begin{bmatrix} -A^{-1}(\partial A/\partial x_j)A^{-1} \\ \hline 0 \end{bmatrix}$$

$$= -\tilde{J}_g^-(\partial J_g/\partial x_j)\tilde{J}_g^- ,$$

we have

$$S_g = -\tilde{J}_g^- R_g \tilde{J}_g^- . \tag{68}$$

On the other hand, we have

$$J_x = (I-\tilde{J}_g^- J_g)D_G . \tag{69}$$

Differentiating it with respect to x we have

$$T_g = -S_g J_g D_G - \tilde{J}_g^- R_g D_G + P_G Z_g . \tag{70}$$

Substituting (68) into (70), we have

$$T_g = -\tilde{J}_g^- R_g (I-\tilde{J}_g^- J_g)D_G + P_G Z_g$$

$$= -\tilde{J}_g^- R_g J_x + P_G Z_g . \tag{71}$$

Since $\Lambda = (\tilde{J}_g^-)^t \nabla f$

and $R_g^t \Lambda = \sum_{i=1}^{m} \lambda_i H_{g_i}$,

we have (66) by substituting (71) into (67). □

Lemma 5.4 A critical point $\bar{x} \in C_f$ of f_v that satisfies rank $J_g(\bar{x}) = m$, is regular if and only if the $m \times m$ matrix

$$K^t(H_f(\bar{x}) - \sum_{i=1}^{m} \lambda_i(\bar{x})H_{g_i}(\bar{x}))K \tag{72}$$

is nonsingular for a n × n matrix K such that

$$J_g(\bar{x})K = 0, \tag{73}$$

where

$$\Lambda(\bar{x}) = (\lambda_i(\bar{x})) = (\tilde{J}_g^-(x))^t \nabla_x f(\bar{x}). \tag{74}$$

Proof Let $\xi = (\xi_j)$, $j = 1, 2, \ldots, n - m$, be local coordinate system of V_g at $\bar{x}$ and let K be defined by $K \equiv (\partial x_i(\xi)/\partial \xi_j)$, then rank K = m. In some neighborhood $U \subset V_g$ of $\bar{x}$, we have

$$g(x(\xi)) = 0 \text{ for } x(\xi) \in U. \tag{75}$$

Differentiating both sides of (75), we have (73). Differentiating them again we have

$$\sum_{i=1}^{n} (\partial x_i/\partial \xi_j)(\partial J_g/\partial x_i)K + J_g(\partial K/\partial \xi_j) = 0 \tag{76}$$

for $j = 1, 2, \ldots, n - m$. On the other hand, we have

$$(\partial f_v/\partial \xi_i) = (\partial f(x(\xi))/\partial \xi_i) = K^t \nabla f.$$

Differentiating both sides of this equality again, we have

$$H_{f_v} \equiv (\partial^2 f_v/\partial \xi_i \partial \xi_j) = [\,(\partial K^t/\partial \xi_i)\nabla f : \ldots : (\partial K^t/\partial \xi_n)\nabla f]$$

$$+ K^t H_f K. \tag{77}$$

From (76) we have

$$\partial K/\partial \xi_j = -\tilde{J}_g^- \left(\sum_{i=1}^{n} (\partial x_i/\partial \xi_j)\,(\partial J_g/\partial x_i)K\right)$$

$$+ (I - \tilde{J}_g^- J_g)R_j, \tag{78}$$

where R_j is some mapping from R^n to R^n. Substituting (78) into (77) we have

$$H_{f_v} = K^t (H_f K - \sum_{i=1}^{n} (\partial x_i/\partial \xi_j)(\partial J_g^t/\partial x_i)\,(\tilde{J}_g^-)^t \nabla f)$$

$$+ R_j^t (I - \tilde{J}_g^- J_g)\nabla f$$

$$= K^t (H_f - \sum_{i=1}^{m} \lambda_i H_{g_i})K, \tag{79}$$

since $(I - \tilde{J}_g^- J_g)\nabla f = 0$ at the critical point $\bar{x}$ of f_v. □

By this lemma we have the well-known second order sufficient condition for an isolated local maximum of f_v.

Lemma 5.5 Under the same assumption, a regular critical point $\bar{x}$ is a maximal point of f_v, if and only if

$$y^t K^t (H_f(\bar{x}) - \sum_{i=1}^{m} \lambda_i(\bar{x}) H_{g_i}(\bar{x}))Ky < 0 \text{ for } y \neq 0,$$

where K is an $n \times m$ matrix which satisfies (73).

Proof The result is a direct consequence of the previous lemma. A different proof is given, for example, in [30]. □

Now we can prove Theorem 5.2. Since the last three terms in (66) vanish at the critical point $\bar{x} \in C_f$, we have

$$J_{\Psi_G}(\bar{x}) = -(I-P_G) + P_G D_G D_G^t P_G^t H \tag{80}$$

where

$$P_G = I - \tilde{J}_g^-(\bar{x}) J_g(\bar{x}), \tag{81}$$

$$D_G = D_G(\bar{x}) \tag{82}$$

and
$$H = H_f(\bar{x}) - \sum_{i=1}^{m} \lambda_i(\bar{x}) H_{g_i}(\bar{x}) . \tag{83}$$

Noting that

$$K = J_x = P_G D_G,$$

$(P_G D_G)^t H (P_G D_G)$ has $(n - m)$ negative eigenvalues

$$-\mu_{m+1}, -\mu_{m+2}, \dots , -\mu_n, (\mu_i > 0) \tag{84}$$

by the previous two lemmas. Hence the matrix $P_G D_G D_G^t P_G^t H$ has the same $n - m$ negative eigenvalues (84) and m zero eigenvalues, because $P_G D_G D_G^t P_G^t H = KK^t H$ and $K^t HK$ have identical sets of nonzero eigenvalues [3]. The projector $(I-P_G)$ has m eigenvalues which are equal to one and $n - m$ eigenvalues which are zero. Let the independent eigenvectors (or principal vectors) of PH, which correspond to the eigenvalues (84), be $\{q_{m+1}, q_{m+2}, \dots , q_n\}$ and let the independent eigenvectors of P, which correspond to the zero eigenvalue, be $\{q_1, q_2, \ldots, q_m\}$ then

$$J_{\Psi_G}(\bar{x})Q = QJ^* \tag{85}$$

where the matrices Q and J* are defined respectively by

$$Q = [q_1 : q_2 : \ldots : q_n],$$

and

$$J^* = \left[\begin{array}{cccc|cccc} -i & & & & & & & \\ & -i & & & & & 0 & \\ 0 & & \ddots & & & & & \\ & & & -i & & & & \\ \hline & & & & -\mu_{m+1} & & & \\ & & & & & -\mu_{m+2} & 0 & \\ & & & & & & \ddots & \\ & & * & & & * & & -\mu_n \end{array}\right] . \tag{86}$$

Since $q_1, q_2, \ldots, q_m$ belong to the null space $N(P_G) = R(I-P_G)$ of P_G and $q_{m+1}, q_{m+2}, \ldots, q_n$ belong to the range $R(P_G)$ of P_G, $R(P_G) + N(P_G) = R^n$, Q is nonsingular. Hence $J_{\Psi_G}(\bar{x})$ is similar to J* and has n negative eigenvalues $-1, -1, \ldots, -1, -\mu_{m+1}, -\mu_{m+2}, \ldots, -\mu_n$. Thus by the Liapunov's direct method, the system (59) is asymptotically stable at the regular maximal point $\bar{x}$.

From the foregoing analysis, it is seen that the autonomous system (59) defines a vector field in U - S and that the set of maximal points are the attractors of the vector field. If x is infeasible, i.e., $x \notin V_g$, then the effect of the correction term

$$-\tilde{J}_g^-(x)g(x)$$

is generally dominant over the gradient projection term,

$$(I-\tilde{J}_g^-(x)J_g(x))\nabla f(x),$$

in the right-hand side $\Psi_G(x)$ of (59). The trajectories of the solutions of the system are first pulled towards the feasible set V_g and then move along the feasible set towards a point with greater $f(x)$ value.

When the Euler method is applied to the system (59), the numerical method behaves, in the initial iterations, like a modified Newton-Raphson method for restoring violation of constrained equations, and in subsequent iterations, the nature of the generalized reduced gradient method becomes more pronounced.

So far we have not specified $D_G(x)$ in (59) (or (26)) except for the case where $D(x) = I_{n-m}$ and $\pi = I$ which amounts to a continuous analogue of the GRG method in (26). Here we suggest another choice for $D_G(x)$, which uses the second order derivative information about $f(x)$ and the $g_i(x)$'s. Let $L(x)$ be an $n \times (n-m)$ matrix defined by

$$P_I(x) \equiv \begin{bmatrix} -A^{-1}(x)B(x) \\ \hline I_{n-m} \end{bmatrix},$$

where $A(x)$ and $B(x)$ are the same matrices as used in (21), in which case $P_G(x) = \pi[0 : P_I(x)]\,\pi^t$. Let $H(x) \equiv H_f(x) - \sum_{i=1}^{m} \lambda_i(x) H_{g_i}(x)$. We put

$$M(x) \equiv -P_G^t(x)H(x)P_G(x)$$

in (58), which amounts to putting

$$D_G(x)D_G^t(x) \equiv M^{\dagger}(x) = -\pi \left[\begin{array}{c|c} 0 & 0 \\ \hline 0 & (P_I{}^t(x)\pi^t H(x)\pi P_I(x))^{-1} \end{array}\right] \pi^t$$

in (59), where $M^{\dagger}(x)$ is the Moore-Penrose inverse of $M(x)$, in which case we have

$$\frac{dx}{dt} = \Psi_G^M(x) = -\tilde{J}_g^-(x)g(x) + P_G(x)M^{\dagger}(x)P_G^t(x)\nabla f(x). \tag{87}$$

This choice for $D_G(x)$ seems to be desirable, because then all the eigenvalues of the Jacobian matrix,

$$J_{\Psi_G}(\bar{x}) = -(I-P_G(\bar{x})) + P_G(\bar{x})M^{\dagger}(\bar{x})P_G^t(\bar{x})H(\bar{x})$$

of $\Psi_G^M(x)$ at a regular optimal point $\bar{x}$ are -1, which can be easily seen by noting that $P_G(\bar{x})M^{\dagger}(\bar{x})P_G^t(\bar{x})H(\bar{x})$ and $M^{\dagger}(\bar{x})P_G(\bar{x})H(\bar{x})P_G(\bar{x}) = M^{\dagger}(\bar{x})M(\bar{x})$ have an identical set of nonzero iegenvalues and that $M(\bar{x})$ is a nonpositive definite symmetric matrix and $P_I^t(x)\pi^t H(x)\pi P_I(x)$ is an $(n-m)\times(n-m)$ negative definite matrix by Lemma 5.5.

Various algorithms can be obtained by choosing different numerical intergration methods for the systems (59), (60) and (87). For the same reason as is given in Section 1 concerning the previous method, higher order intergration methods are preferable to the Euler method. However, the purpose is not to pursue the trajectory very accurately, but to trace the curved surface of the feasible set V_g rather crudely, approaching a maximal point. Hence the absolute stability is of uppermost importance in the choice of numerical intergration methods, since the Jacobian matrix $J_{\Psi_G}(\bar{x})$ given in (80) is similar to (86) at a regular optimal point $\bar{x}$. Methods with larger interval of absolute stability [29] are desirable for our purpose. If a method with the absolute stability interval $(-\omega,0)$ is applied to the system (59) (or (60)) and if the initial point is close enough to a regular maximal point $\bar{x}$, the numerical solution converges to the maximal point as long as the step-sizes (Δt) are kept so small as to satisfy the inequality, $0 < \mu_j(\Delta t) < \omega$, for $j = 1,2,\ldots,n$ where we use a convention $\mu_1 = \mu_2 = \ldots = 1$, $\mu_{m+1}, \mu_{m+2}, \ldots, \mu_n$ are given in (84). In the case of (87), the step-sizes (Δt) should be chosen to satisfy the inequality $0 < \nabla t < \omega$.

Since a lengthy process is required for evaluating $\Psi(x)$, the explicit methods should be chosen as the intergration method for (59) (or (60), (87)). A few algorithms were described in the previous paper [49] for the numerical intergration of these systems.

6. QUASI-NEWTON ALGORITHM AND MODIFIED NEWTON ALGORITHM

Although our method has a fairly large domain of attraction to a maximal point of f_v and the speed of the convergence is faster than one might think, the methods do not generally have the superlinear convergence property in a neighborhood of a maximal point. So if the number of variables is not large, it might be better to switch the method to one which has that property when an approximation becomes close enough to the solution. We will describe two algorithms for this purpose. One method is obtained by applying a quasi-Newton method to the system of nonlinear equations,

$$\Psi(x) \equiv J_g^-(x)g(x) + (I-J_g^-(x)J_g(x))\nabla f(x) = 0, \tag{88}$$

and the other by applying a modified Newton-Raphson method. Note here that under assumption (4) the condition (88) is equivalent by (54) to the first order necessary conditions,

$$\nabla f(x) = J_g^t(x)\Lambda(x) \quad \text{and} \quad g(x) = 0,$$

for optimality.

Since the first derivative (Jacobian) J_Ψ of $\Psi(x)$ is not a symmetric matrix at a critical point of f_v, as is seen in the proof of Theorem 5.2, we apply the Broyden's method [10] to obtain the following algorithm.

QUASI-NEWTON ALGORITHM. Generate a sequence of vectors by the recurrence relation,

$$x^{(k+1)} = x^{(k)} - J^{(k)}\Psi(x^{(k)}), \tag{89}$$

where $J^{(0)}$ is the initial guess of J_Ψ, e.g., $J^{(0)} \doteq -I$,

$$y^{(k)} = \Psi(x^{(k)}) - \Psi(x^{(k-1)}),$$

$$p^{(k)} = x^{(k)} - x^{(k-1)},$$

$$J^{(k+1)} = J^{(k)} - (J^{(k)}y^{(k)}-p^{(k)})\ (p^{(k)})^t J^{(k)}/(p^{(k)})^t J^{(k)} y^{(k)} .$$

Note that this algorithm requires only the first-order derivative information on the functions involved.

Since the Jacobian J_Ψ may be well approximated around a critical point by neglecting the last three terms in (66), we obtain the following algorithm.

MODIFIED NEWTON-RAPHSON ALGORITHM. Generate a sequence of vectors by the recurrence formula,

$$x^{(k+1)} = x^{(k)} - (\bar{J}_\Psi(x^{(k)}))^{-1}\ \Psi(x^{(k)}) \tag{90}$$

$$= -J_g^-(x^{(k)})g(x^{(k)}) + X^-(x^{(k)})P_I(x^{(k)})P_I^t(x^{(k)})$$

$$(H(x^{(k)})J_g^-(x^{(k)})g(x^{(k)}) - \nabla f(x^{(k)})), \tag{91}$$

where

$$\bar{J}_\Psi(x) \equiv -Q(x) + P(x)P^t(x)H(x),$$

$$P(x) \equiv I - J_g^-(x)T_g(x) = [0 \mid P_I(x)],$$

$$Q(x) \equiv I - P(x) = J_g^-(x)J_g(x) = \begin{bmatrix} Q_B(x) \\ \hline 0 \end{bmatrix},$$

$$P_I(x) \equiv \begin{bmatrix} -E(x) \\ \hline I_{n-m} \end{bmatrix},$$

$$Q_B(x) \equiv [I_m \mid E(x)],$$

$$E(x) \equiv (J_g(x))_B^{-1}(J_g(x))_I,$$

$$X(x) \equiv P(x)P^t(x)H(x)P(x) = P(x)[0 \mid P_I^t(x)H(x)P_I(x)],$$

$$X^-(x) \equiv P(x)[0 \mid (P_I^t(x)H(x)P_I(x))^{-1}],$$

$$H(x) \equiv H_f(x) - \sum_{i=1}^{m} \lambda_i(x)H_{g_i}(x),$$

and

$$\Lambda(x) \equiv (\lambda_i(x)) = (J_g^-(x))^t \nabla f(x).$$

Lemma 6.1 The formulas (90) and (91) are equivalent in a neighborhood of a regular critical point of f_v.

Proof We will drop the argument x in the following expressions. Let $Y = PP^tHQ$, then we have $(-Q+X^-Y+X^-)(-Q+PP^tH) = Q-X^-Q+X^-X = Q+P = I$ since $PP^tH = X+Y$, $X^-YQ = X^-Y$, $QP = 0$, $YP = 0$, $X^-Q = 0$ and $X^-X = P$. Thus

$$(\bar{J}_\Psi)^{-1} = (-Q + PP^tH)^{-1} = Q + X^-Y + X^-. \tag{92}$$

Substituting (92) into (90) we have the desired result since $QJ_g^- = J_g^-$, $QP = 0$, $YP = 0$ and $X^-J_g^- = 0$. □

Theorem 6.2 The rate of convergence of the algorithm in a neighborhood of a regular critical point $\bar{x}$ of f_v is quadratic.

Proof Following Ortega-Rheinboldt [37], the iteration (90) is expressed in the form

$$x^{(k+1)} = G(x^{(k)})$$

where $G(x) \equiv x - (\bar{J}_\Psi(x))^{-1}\Psi(x)$. $G(x)$ is differentiable at $\bar{x}$ and its Jacobian J_G at $\bar{x}$ is given by

$$J_G(\bar{x}) = I - (\bar{J}_\Psi(\bar{x}))^{-1}J_\Psi(\bar{x}) = 0,$$

since $\Psi(\bar{x}) = 0$ and $\bar{J}_\Psi(\bar{x}) = J_\Psi(\bar{x})$. We have

$$G(x) - G(\bar{x}) = x - \bar{x} - (\bar{J}_\Psi(x))^{-1}\Psi(x), \tag{93}$$

and

$$\Psi(x) = J_\Psi(x)\ (x - \bar{x}) + 0\ (||x - \bar{x}||^2). \tag{94}$$

By (66), we have $J_\Psi(x) = \bar{J}_\Psi(x) + R(x)$, and $R(\bar{x}) = 0$. In a neighborhood of $\bar{x}$, we have $R(x) = 0\ (||x - \bar{x}||)$. Hence

$$J_\Psi(x) = \bar{J}_\Psi(x) + 0\ (||x - \bar{x}||). \tag{95}$$

Substituting (94) and (95) into (93), we have $G(x) - G(\bar{x}) = 0\ (||x - \bar{x}||^2)$. Thus by Theorem 10.1.6 of Ortega-Rheinboldt [37], we have the desired result. □

REFERENCES

1 J. Abadie, and J. Carpentier, Generalization of the Wolfe Reduced Gradient Method to the Case of Nonlinear Constraints, Optimization, Edited by R. Fletcher, Academic Press, London, England, 1969. pp. 37-47.

2 R. Abraham and J. Robbin, Transversal Mapping and Flows, Benjamin, New York, 1967.

3 S. N. Afriat, Orthogonal and Oblique Projectors and the Characteristics of Pairs of Vector Spaces, Proceeding of Cambridge Philosophical Society, Vol. 53, pp. 800-816, 1957.

4 L. Auslander and R. E. MacKenzie, Introduction to Differentiable Manifolds, McGraw-Hill, New York, 1963.

5 Y. Bard and J. L. Greenstadt, A Modified Newton Method for Optimization with Equality Constraints, Optimization, Edited by R. Fletcher, Academic Press, London, England, 1968.

6 D. P. Bertsekas, Combined Primal-Dual Penalty Methods for Constrained Minimization, SIAM Journal on Control, Vol. 13, pp. 521-544, 1975.

7 D. P. Bertsekas, Multiplier Methods: A Survey, Automatica, Vol. 12, pp. 133-145, 1976.

8 N. P. Bhatia and G. P. Szego, Dynamical Systems: Stability Theory and Applications, Springer Verlag, Berlin, New York, 1967.

9 F. H. Branin, Widely Convergent Method for Finding Multiple Solutions of Simultaneous Nonlinear Equations, IBM Journal of Research and Development, September, pp. 504-521, 1972.

10 C. G. Broyden, A Class of Methods for Solving Nonlinear Simultaneous Equations, Mathematics of Computation, Vol. 19, pp. 577-593, 1965.

11 C. W. Carroll, The Created Response Surface Technique for Optimizing Nonlinear Restricted Systems, Operations Research, Vol. 9, pp. 169, 1961.

12 J. E. Dennis and J. J. More, A Characterization of Superlinear Convergence and its Application to Quasi-Newton Methods, Mathematics of Computation, Vol. 28, pp. 549-560, 1974.

13 A. V. Fiacco and G. P. McCormick, Nonlinear Programming: Sequential Unconstrained Minimization Techniques, John Wiley and Sons, New York, 1968.

14 R. Fletcher, A Class of Methods for Nonlinear Programming with Termination and Convergence Properties, Integer and Nonlinear Programming, Edited by J. Abadie, North Holland, Amsterdam, pp. 157-175, 1970.

15 R. Fletcher and S. A. Lill, A Class of Methods for Nonlinear Programming II; Computational Experience, Nonlinear Programming, Edited by J. B. Rosen et. al., Academic Press, New York, pp. 67-92, 1970.

16 R. Fletcher, A Class of Methods for Nonlinear Programming III: Rate of Convergence, Numerical Methods for Nonlinear Optimization, Edited by F. A. Lootsma, Academic Press, London, New York, pp. 371-381, 1972.

17 R. Fletcher, An Exact Penalty Function for Nonlinear Programming with Inequalities, Mathematical Programming, Vol. 5, pp. 129-150, 1973.

18 R. Fletcher, Methods Related to Langrangian Functions, Numerical Methods for Constrained Optimization, R. E. Gill and W. Murray (eds), Academic Press, London, 219-239, 1974.

19 P. E. Gill and W. Murray, Newton-Type Methods for Linearly Constrained Optimization, Numerical Methods for Constrained Optimization, P. E. Gill and W. Murray (eds), Academic Press, London, 29-66, 1974.

20 P. E. Gill, G. H. Golub, W. Murray, and M. A. Saunders, Methods for Modifying Matrix Factorizations, Mathematics of Computation, Vol. 28, pp. 505-535, 1974.

21 W. A. Gruver and N. H. Engersbach, A Variable Metric Algorithm for Constrained Minimization Based on an Augmented Lagrangian, International Journal for Numerical Methods in Engineering, Vol. 10, pp. 1047-1056, 1976.

22 W. A. Gruver and K. Tanabe, Continuous Analogs of the Gradient Projection Method in Nonlinear Programming, 1977 IEEE Conference on Decision and Control Proceedings, 1977.

23 H. Hancock, Theory of Maxima and Minima, Gin and Company, 1917, and Dover Publications, 1960.

24 M. R. Hestenes, Multiplier and Gradient Methods, Journal of Optimization Theory and Applications, Vol. 4, pp. 303-320, 1969.

25 Y. H. Huang and S. Naqvi, Unconstrained Approach to the Extremization of Constrained Functions, Journal of Mathematical Analysis and Applications, Vol. 39, pp. 360-374, 1972.

26 H. Y. Huang and A. K. Aggarwal, A Class of Quadratically Convergent Algorithms for Constrained Function Minimization, Journal of Optimization Theory and Applications, Vol. 16, pp. 447-485, 1975.

27 H. J. Kelley, W. F. Denham, I. L. Johnson, and P. O. Wheatley, An Accelerated Gradient Method for Parameter Optimization with Nonlinear Constraints, The Journal of the Astronautical Science, Vol. 13, pp. 166-169, 1966.

28 H. Kwakernaak and R. C. W. Strijbos, Extremization of Functions with Equality Constraints, Mathematical Programming, Vol. 2, pp. 279-295, 1972.

29 J. D. Lambert, Computational Methods in Ordinary Differential Equations, John Wiley and Sons, London, New York, Sydney, 1973.

30 O. L. Mangasarian, Nonlinear Programming, McGraw-Hill, New York, New York, 1968.

31 Y. Matsushima, Differentiable Manifolds, Marcel Dekkar Inc., New York, 1972.

32 K. Mortensson, A New Approach to Constrained Function Optimization. Journal of Optimization Theory and Applications, Vol. 12, pp. 531-554, 1973.

33 G. P. McCormick, A Second Order Method for the Linearly Constrained Nonlinear Programming Problem, Nonlinear Programming, J. B. Rosen, O. L. Mangasarian and K. Ritter (eds), Academic Press, New York, 207-243, 1970.

34 G. P. McCormick, Penalty Function Versus Non-penalty Function Methods for Constrained Nonlinear Programming Problems, Mathematical Programming, Vol. 1, pp. 217-238, 1971.

35 A. Miele, H. Y. Huang, and J. C. Heideman, Sequential Gradient-Restration Algorithm for the Minimization of Constrained Functions, Ordinary and Conjugate Gradient Versions. Journal of Optimization Theory and Applications, Vol. 4, pp. 213-243, 1969.

36 A. Miele, A. V. Levy and E. E. Cragg, Modifications and Extensions of the Conjugate-Restration Algorithm for Mathematical Programming Problems, Journal of Optimization Theory and Applications, Vol. 7, pp. 450-472, 1971.

37 J. M. Ortega and W. C. Rheinboldt, Iterative Solution of Nonlinear Equations in Several Variables, Academic Press, New York, London, 1970.

38 M. J. D. Powell, A Method for Nonlinear Constraints in Minimization Problems, Optimization, Edited by R. Fletcher, Academic Press, London, England, 1969.

39 M. J. D. Powell, Algorithms for Nonlinear Constraints that Use Lagrangian Functions, Mathematical Programming 14, 224-248, 1978.

40 J. O. Ramsay, A Family of Gradient Methods for Optimization: The Computer Journal, Vol. 13, 413-417, 1970.

41 R. T. Rockafellar, The Multiplier Method of Hestenes and Powell Applied to Convex Programming, Journal of Optimization Theory and Applications, Vol. 12, 555-562, 1973.

42 J. B. Rosen, The Gradient Projection Method for Nonlinear Programming. Part II: Nonlinear Constraints, Journal of Society of Industrial and Applied Mathematics, Vol. 9, 514-532, 1961.

43 R. D. Rupp, On the Combination of the Multiplier Method of Hestenes and Powell with Newton's Method, Journal of Optimization Theory and Applications, Vol. 15, 167-187, 1975.

44 R. W. H. Sargent, Reduced-gradient and Projection Methods for Nonlinear Programming, Numerical Methods for Constrained Optimization, R. E. Gill and W. Murray (eds), Academic Press, London, 149-174, 1974.

45 K. Tanabe, Algorithm for Constrained Maximization Technique for Nonlinear Programming, Proceeding of the Second Hawaii International Conference on System Sciences, University of Hawaii, 487-490, 1969.

46 K. Tanabe, An Algorithm for Constrained Maximization in Nonlinear Programming, Institute of Statistical Mathematics, Research Memorandum No. 31, 1969.

47 K. Tanabe, An Algorithm for the Constrained Maximization in Nonlinear Programming, Journal of Operations Research Society of Japan, Vol. 17, 184-201, 1974.

48 K. Tanabe, Analog-type Nonlinear Programming (in Japanese), Mathematical Sciences, Vol. 157, 45-51, 1976. Presented at the Japan-France Seminar on Functional Analysis and Numerical Analysis, Tokyo, 1976.

49 K. Tanabe, A Geometric Method in Nonlinear Programming, Stanford University Computer Science Report, STAN-CS-77-643, 1977. Presented at the Gatlinburg VII Conference, Asilomar, 1977.

50 K. Tanabe, Continuous Newton-Raphson Method for Solving an Underdetermined System on Nonlinear Equations, BNL Report #23573, AMD 784, 1977.

51 R. A. Tapia, Diagonalized Multiplier Methods and Quasi-Newton Methods for Constrained Optimization, Journal of Optimization Theory and Applications, Vol. 22, 135-194, 1977.

52 R. A. Tapia, Quasi-Newton Methods for Equality Constrained Optimization: Equivalence of Existing Methods and a New Implementation, Manuscript, Presented at Nonlinear Programming Symposium 3, Madison, Wisconsin, July 1977.

53 M. A. H. Wright, Numerical Methods for Nonlinearly Constrained Optimization, Stanford University SLAC Report, No. 193, March 1976.

54 H. Yamashita, Differential Equation Approach to Nonlinear Programming, Computation and Analysis, Vol. 7, 11-38, 1976.

55 G. Zoutendijk, Method of Feasible Directions; A Survey in Linear and Nonlinear Programming, American Elseview Pub. Co., Amsterdam, New York, 1960.

This work was done while the author was a visitor at the Applied Mathematics Department, Brookhaven National Laboratory, between September 1977 and December 1978. He is very much pleased to acknowledge the gracious hospitality and stimulating environment provided by Dr. Robert Marr, Dr. Charles Goldstein, and Dr. James Ho, at Brookhaven National Laboratory.

Kunio Tanabe
The Institute of Statistical Mathematics
Minamiazubu
Minatoku, Tokyo

and

Applied Mathematics Department
Brookhaven National Laboratory
Upton, New York

4 Ring theoretic extensions

M P DRAZIN

Differentiation of generalized inverses

ABSTRACT

Golub and Pereyra [4] and Decell [2] have obtained an explicit formula for the derivative of the Moore-Penrose inverse of an arbitrary differentiable matrix-valued function, while more recently, Campbell [1] and Hartwig and Shoaf [5] have done the same for the pseudo-inverse introduced by the author [3].

Simple alternative proofs are now presented, which have the advantage of giving some information for other and more general types of inverses. It is shown also that, if a continuous $n \times n$ matrix-valued function $A(t)$ has <u>any</u> type of continuous generalized inverse on a connected open set U, then A must be either invertible throughout U or non-invertible throughout U.

1. DIFFERENTIATION OF (2)-INVERSES

For any square complex matrix A, both the Moore-Penrose inverse $X = A^{\dagger}$ of A and the author's pseudo-inverse $X = A'$ of A share the familiar "(2)-inverse" property $X = XAX$; more generally, we shall use the same terminology for elements a, x in any associative ring Q. Obviously $x = 0$ is a (2)-inverse for every $a \in Q$ (and a given nonzero $a \in Q$ may or may not have other (2)-inverses). In the matrix case, the definitions of the pseudo-inverse and of the Moore-Penrose inverse each automatically determine a *unique* x corresponding to each given a; however, in our arguments in this section, we shall not need to assume uniqueness in this sense. Instead, we shall (at first) think of $a = a(t)$ as being a given function (taking its values in Q) of some independent variable t, and of $x = x(t)$ as being *any* function (again, with values in Q) of t subject to "differentiability" and the (2)-inverse requirement that $x(t) = x(t)a(t)x(t)$ for all t near some specified point t_o. More precisely, if t varies in a given topological space T, then a, x are maps $a,x:T \to Q$ such that $x = xax$ throughout some neighborhood U of t_o.

Let R denote the set of all maps from U into Q (so that, by restriction, we may regard a,x as elements of R), regarded as a ring in the usual way. In order to give a concrete meaning to the differentiability (e.g. in the Fréchet sense) of a and x, it would be necessary to assume a metric structure in U and Q; but in fact our arguments will not use even the fact that T has a topology (indeed, T will not be mentioned again). All we shall need is that there be given some operator d, applicable to each of a, x, ax and xa, which "behaves like differentiation." Specifically, for any ring R and any subset S of R, we shall call a map $d:S \to R$ a <u>derivation on</u> S iff

$$d(yz) = (dy)z + y(dz) \quad \text{whenever} \quad y,\ z,\ yz \in S.$$

(Note that, for our purposes, a derivation need not be additive.) To simplify notation in the course of computations, we shall also write y' as an alternative to dy (this use of the symbolism y' should be distinguished from the use of primes to denote pseudo-inverses as in [3]). We note also that, in contrast to the argument of Campbell, we shall, in most of this section, have no need to assume that

$$a = axa \tag{1}$$

<u>Lemma 1.</u> In any associative ring R, let $a,x \in R$, let d be a given derivation on the set $\{a,x,ax,xa\}$, and write

$$c = dx + x(da)x, \quad e = ax, \quad f = xa.$$

Then, if $$x = xax \tag{2}$$
it follows that $$c = ce + fc \tag{3}$$
and $$fce = 0. \tag{4}$$

<u>Proof</u> (3). On "differentiating" (2), we have

$$x' = x'ax + xa'x + xax' \quad . \tag{5}$$

Also $ce + fc = (x'+xa'x)ax + xa(x'+xa'x) = x'ax + xa'(xax) + xax' + (xax)a'x$. Hence by (2) and (5), we find $ce + fc = x'ax + 2xa'x + xax' = x' + xa'x = c$.

(4). By (3) and (2), we have $fce = (c-ce)e = c(e-e^2) = 0$. $\square$

Corollary 1 Under the hypotheses of Lemma 1, there exist elements p, q of R such that

$$dx = -x(da)x + xp(1-ax) + (1-xa)qx.$$

Proof This follows at once from (3) and (4) with $p = ac$, $q = ca$. □

Corollary 1 may be regarded as a sort of precursor of each of the more specific formulae of Campbell-Hartwig-Shoaf and of Decell-Golub-Pereyra, but is apparently too weak to be of much practical use in itself: we need p, q to be given explicitly, and in terms of a, x, a' alone (i.e. p, q should not involve x'). However, even from Corollary 1, at least it is clear that $x = 0$ implies $dx = 0$.

Lemma 2 If in Lemma 1 we have also

$$e = ax = xa = f, \tag{6}$$

then

$$ecx = xce = 0, \tag{7}$$

$$ec = xca + x^2(da)(1-e), \tag{8}$$

and

$$ce = acx + (1-e)(da)x^2 \quad . \tag{9}$$

Proof (7). By (2) and (6), we have $x = (xa)x = (ax)x = ex$, whence $ecx = fcex = 0$ by (4); similarly $xce = 0$.

(8). Adopting the commutator notation $[r,s] = rs - sr$, we have $[a,c] = [a,x' + xa'x] = [a,x'] + [a,xa'x]$. Also, by differentiating (6), we have $a'x + ax' = x'a + xa'$, i.e. $[a,x'] = [x,a']$. Hence $x[a,c] = x[x,a'] + x[a,xa'x] = x(xa' - a'x) + x(axa'x - xa'xa) = x^2a' - x^2a'xa$ by (2), and this is (8).

(9) follows by symmetry. □

Theorem 1 In any ring R, let a,x be elements satisfying $x = xax$, $ax = xa$, let d be any given derivation on the set $\{a,x,ax,xa\}$, and write $c = dx + x(da)x$. Then

$$dx = -x(da)x + a^kcx^k + x^kca^k$$

$$+ \sum_{j=0}^{k-1} (a^j(1-ax)(da)x^{j+2} + x^{j+2}(da)(1-xa)a^j) \quad (k = 1,2,\ldots) \quad .$$

Proof Substituting from (8) and (9) into (3), we find $c = acx + (1-e)a'x^2 + xca + x^2a'(1-e)$, which we rewrite as

$$c = acx + xca + v + w \quad , \tag{10}$$

where $v = (1-e)a'x^2$, $w = x^2a'(1-e)$. (The equation (10) is a generalization of Theorem 1 of Campbell [1], who proved it subject to the additional hypothesis $a = axa$, in which case, by (7), the terms acx and xca both vanish.) Next, since $x = xax = xe = ex$, clearly

$$xv = wx = 0. \tag{11}$$

By substituting from (10) into itself, we have

$$\begin{aligned} c &= a(acx + xca + v + w)x + x(acx + xca + v + w)a + v + w \\ &= (a^2cx^2 + axce + avx + awx) + (ecxa + x^2ca^2 + xva + xwa) + v + w \\ &= (a^2cx^2 + avx) + (x^2ca^2 + xwa) + v + w \end{aligned}$$

by (7) and (11), i.e. $c = a^2cx^2 + x^2ca^2 + (v+avx) + (w+xwa)$.
Similarly, by a straightforward induction, we obtain

$$c = a^kcx^k + x^kca^k + \sum_{j=0}^{k-1} (a^jvx^j + x^jwa^j) \quad (k = 1,2,\ldots).$$

as required. □

We shall show in Section 2 below that the Campbell-Hartwig-Shoaf formula for the derivative of the pseudo-inverse follows immediately from Theorem 1. Similarly, as preparation for our alternative proof of the Decell-Golub-Pereyra result, we note two trivial lemmas (neither involving any *-operation):

Lemma 3 Under the hypotheses of Lemma 1, we have

$$dx = -x(da)x + x(de) + (df)x.$$

Proof $-xa'x + xe' + f'x = -xa'x + x(a'x + ax') + (x'a + xa')x$
$= xax' + x'ax + xa'x = (xax)' = x'$. □

Lemma 4 For any ring R, any $a, x \in R$ and any derivation d on the set $\{a,x,ax,xa\}$, if

$$a = axa \tag{1}$$

and if we write $e = ax$ and $f = xa$ as above, then

$$(de)e = (1-e)(da)x, \quad f(df) = x(da)(1-f) \quad .$$

Proof Writing (1) in the form $a = af$ and differentiating, we have $a' = a'f + af'$, so that $ff' = xaf' = x(a' - a'f) = xa'(1-f)$. Similarly, $e'e = (1-e)a'x$. □

As a consequence of Lemma 4, obviously $e(de)e = 0 = f(df)f$, and this fact, in the special case where $a^2 = a = x$ (so that $e = f = a$), was first noted by Hearon and Evans [6, Lemma 3] and later utilized by Decell [2].

In Corollary 1, Lemma 2, Theorem 1, and Lemma 4 above (as also in Theorems 2 and 3 below) we have tacitly assumed the presence of a unity element 1 in R; however, this assumption can easily be avoided by appropriately rewriting all expressions in which 1 appears, e.g. $(1-e)a'x = a'x - ea'x$.

2. PROOFS OF THE CAMPBELL-HARTWIG-SHOAF AND DECELL-GOLUB-PEREYRA FORMULAE

We have stated Theorem 1 explicitly so as to emphasize its validity for any (differentiable) a, x satisfying $x = xax$ and $ax = xa$, i.e. without having to assume enough to force x to be the pseudo-inverse of a; similarly, Lemmas 3 and 4 give information about any (differentiable) a, x satisfying $x = xax$ and $a = axa$ without any assumptions corresponding to the requirement that ax and xa be hermitian as in Moore-Penrose theory. However, the principal cases of interest are naturally those discussed respectively by Campbell-Hartwig-Shoaf and by Decell-Golub-Pereyra, and we can now apply Theorem 1 and Lemmas 2, 3, and 4 to re-derive their results rather simply:

Theorem 2 (Campbell-Hartwig-Shoaf). In any associative ring R, let $a \in R$, suppose that a is pseudo-invertible with pseudo-inverse x, and let d be a given derivation on the set $\{a, x, ax, xa\}$. Then, for all sufficiently large k, we have

$$dx = -x(da)x + x^2\left(\sum_{j=0}^{k} x^j(da)a^j\right)(1-ax) + (1-xa)\left(\sum_{j=0}^{k} a^j(da)x^j\right)x^2.$$

Proof By the definition [3] of the pseudo-inverse, we have (2), (6), and also $a^m = a^{m+1}x$ for all sufficiently large m, say for $m \geq m_o$. Then, for $k \geq m_o$, we have $a^k cx^k = a^{k+1}xcx^k = a^k(ecx)x^{k-1} = 0$ by (7), and similarly $x^k ca^k = 0$.

Also $a^j(1-ax) = (1-xa)a^j$ for all j, while $a^j(1-ax) = a^j - a^{j+1}x = 0$

for $j \geq m_o$, so the stated identities now follow from Theorem 1. □

<u>Theorem 3</u> (Decell-Golub-Pereyra). Let R be any associative ring with a given (not necessarily involutory) anti-endomorphism $y \to y^*$, let $a, x \in R$ and assume that

$$a = axa, \tag{1}$$

$$x = xax, \tag{2}$$

and $(ax)^* = ax, \quad (xa)^* = xa$ (12)

(as are the case when $x = a^\dagger$ in Moore-Penrose theory).

Then, for any derivation d which acts on the set $\{a,x,ax,xa\}$ and commutes with the given anti-endomorphism $*$, we have

$$dx = -x(da)x + xx^*(da)^*(1-ax) + (1-xa)(da)^*x^*x.$$

<u>Proof</u> By (12) we have $e^* = e$, so that $(e'e)^* = e^*e'^* = ee^{*\prime} = ee'$. Hence, by Lemma 4, $xe' = (xax)e' = x(ax)e' = xee' = x(e'e)^* = x((1-e)a'x)^*$ $= xx^*a'^*(1^*-e^*) = xx^*a'^*(1-ax)$ and similarly $f'x = (1-xa)a'^*x^*x$. The desired identity now follows from Lemma 3. □

3. CONTINUOUS GENERALIZED INVERSES

Hearon and Evans [6, Theorem 1] have shown that, if A, X are continuous matrix functions satisfying $A = AXA$ on an interval, then A must have constant rank on the interval. The hypothesis $A = AXA$ applies when X is the Moore-Penrose inverse of A, but not (in general) when X is the pseudo-inverse. However, there is an analogous result which applies in both cases. To state this succinctly, we introduce a binary relation between matrices: given $n \times n$ matrices A, X, let us call X an <u>admissible inverse</u> for A iff either A has no inverse or $AX = XA = I$ (where I denotes the $n \times n$ identity matrix). For any A, clearly this property holds when X is either the Moore-Penrose inverse or the pseudo-inverse of A (or indeed <u>any</u> other well-defined "generalized inverse" which anyone might wish to consider).

<u>Theorem 4</u> If a given $n \times n$ matrix-valued function A of t has an admissible inverse function X near $t = t_o$ such that both A and X are continuous at t_o, then either A is non-singular throughout some neighborhood of t_o or else A is singular throughout some neighborhood of t_o.

Proof If $\det A(t_o) \neq 0$, then, by continuity, $\det A(t)$ is nonzero throughout some neighborhood of t_o.

If $\det A(t_o) = 0$, then $(\det (AX))(t_o) = (\det A(t_o))(\det X(t_o)) = 0$. By the continuity of A, X and hence of AX, $\det AX$, consequently $|\det AX| < 1$, and a fortiori $AX \neq I$, throughout some neighborhood of t_o. But then, by admissibility, we have $\det A=0$ throughout the same neighborhood. □

REFERENCES

1. S. L. Campbell, Differentiation of the Drazin inverse. SIAM J. Appl. Math. 30(1976), 703-707.

2. H. P. Decell, On the derivative of the generalized inverse of a matrix. Linear & Multilin. Alg. 1 (1974), 357-359.

3. M. P. Drazin, Pseudo-inverses in associative rings and semigroups. Amer. Math. Monthly 65(1958), 506-514.

4. G. H. Golub and V. Pereyra, The differentiation of pseudo-inverses and nonlinear least squares problems whose variables separate. SIAM J. Numer. Anal. 10(1973), 413-432.

5. R. E. Hartwig and J. Shoaf, On the derivative of the Drazin inverse of a complex matrix, SIAM J. Math. Anal. 10(1979), 207-216.

6. J. Z. Hearon and J. W. Evans, Differentiable generalized inverses. J. Res. Nat. Bur. Standards 72B (1968), 109-113.

Michael P. Drazin
Department of Mathematics
Purdue University
West Lafayette, Indiana 47907

R E HARTWIG AND F J HALL

Applications of the Drazin inverse to Cesaro-Neumann iterations

ABSTRACT

The convergence behavior and the closed form limit of the Cesaro-Neumann iterations $S_n^k = \sum_{r=0}^{n-1} (I - XA)^r X/n^k$ and $D_n^k = \sum_{r=0}^{n-1} (n - r)(I - XA)^r X/n^{k+1}$ $k = 0,1,\ldots$ are investigated. This investigation is based on a detailed study of the Cesaro sums $C_n^k(B) = \sum_{r=0}^{n-1} B^r/n^k$, $D_n^k(B) = \sum_{r=0}^{n-1} (n - r)B^r/n^{k+1}$ and their algebraic and analytic interaction with governing matrices of the type $U = XAXX^{=} + I - XX^{-}$, where $XX^{-}X = X = XX^{=}X$. It is shown how suitable algebraic identities between $C_n^k(B)$, $D_n^k(B)$ and B^n/n^k, $[B(B - I)(B - I)^{-}]^n/n^k$, can be combined with the spectral theorem to derive closed form expressions for the limit of these types of iterations in terms of Drazin inverses. A study is made of the relation between similarity and inner inverses $(\cdot)^{-}$, and this is used to derive some remarkable spectral properties of matrices of the form $B(B - \gamma I)(B - \gamma I)^{-}$. The algebraic, asymptotic and spectral properties of the governing matrices are examined and their Drazin inverses are calculated. The Cesaro-Neumann iterations are then used to give Neumann-type expansions for outer inverses of a matrix including the Drazin and Moore-Penrose generalized inverses.

1. INTRODUCTION

In a recent paper [21], Tanabe investigated the Neumann-type expansion

$$\sum_{r=0}^{\infty} (I - XA)^r X \tag{1.1}$$

for a reflexive generalized inverse [1] of a complex matrix A. This type of expansion includes many of the iterative schemes used in numerical analysis [22]. In this paper we shall investigate the convergence behavior and the closed form expressions for the limit of this and the more general type of Cesaro-Neumann iterations:

$$S_n^k = \sum_{r=0}^{n-1} (I - XA)^r X/n^k \quad \text{and} \tag{1.2a}$$

$$T_n^k = \sum_{r=0}^{n-1} (n - r)(I - XA)^r X/n^{k+1}, \quad n \to \infty, \tag{1.2b}$$

where k will always denote a non-negative integer.

We shall demonstrate that these iterations are completely determined by the Cesaro sum matrix $C_n^k(I - U_1)$ and the "differentiated" Cesaro sum matrix $D_n^k(I - U_1)$ respectively, where

$$C_n^k(A) = \sum_{r=0}^{n-1} A^r/n^k \tag{1.3a}$$

$$D_n^k(A) = \sum_{r=0}^{n-1} (n - r)A^r/n^{k+1}, \quad k \geq 0 \tag{1.3b}$$

and

$$U_1 = XAXX^- + I - XX^-.$$

The identities

$$S_n^k = [C_n^k(I - U_1)]X, \quad T_n^k = [D_n^k(I - U_1)]X \tag{1.4}$$

will then be used to show that if S_n^k converges to, say, G and T_n^k converges to, say, H, then

$$G = \begin{cases} (XA)^{\#}X & \text{if } k = 0 \\ (-1)^{k-1}[I - XA(XA)^d](XA)^{k-1}X/k! & \text{if } k \geq 1, \end{cases} \tag{1.5a}$$

$$H = \frac{1}{k+1} G, \quad k \geq 0 \tag{1.5b}$$

where $(\cdot)^{\#}$ and $(\cdot)^d$ denote the group and Drazin generalized inverses of $(\cdot)$ respectively [1]. It will further be shown that

$$S_n^k = [C_n^k(I - P)]X = XC_n^k(I - V_1) = XC_n(I - Q) \tag{1.6}$$

$$T_n^k = [D_n^k(I - P)]X = XD_n^k(I - V_1) = XD_n^k(I - Q), \quad \text{where}$$

$$V_1 = X^-XAX + I - X^-X \quad P = XA + I - XX^- \quad \text{and} \quad Q = AX + I - X^-X \tag{1.7}$$

so that there are actually four ways of tackling these problems, depending on which matrix one selects to govern the iteration.

Our main tools in this investigation will be

(i) algebraic identities between $C_n^k(A)$, $D_n^k(A)$ and the matrices A^n/n^k, $A_1 = A(A - I)(A - I)^-$,

(ii) certain master identities involving U_1, V_1, P, Q, their Drazin inverses and XA, AX,

(iii) the spectral theorem for matrices.

This paper ties up many topics in the theory of convergent iterations and is purposely done as algebraic as possible, avoiding canonical forms, with generalizations to rectangular matrices and operators in mind.

This paper is divided into five sections. In the first section we introduce our concepts and notations. In section 2 we furnish some fundamental working lemmas, while in section 3 we give a detailed analysis of the algebraic and spectral properties of Cesaro and differentiated Cesaro sums. In the fourth section the algebraic and spectral properties of the governing matrices U_1, V_1, P and Q are closely examined and in the last section we combine all of these results to investigate the convergence and limit of S_n^k, T_n^k.

Let us begin by defining some of our concepts. Throughout this paper we shall denote the range, the row space, the right nullspace and the left nullspace $\{\tilde{y}^T | \tilde{y}^T A = \tilde{0}^T\}$ of a matrix A by R(A), RS(A), N(A) and °A respectively. We shall always assume (except for (2.1)) that our matrices are over a regular ring $\mathcal{R}$ with unity, for which $a \in a\mathcal{R}a$, for all a in $\mathcal{R}$. This ensures [23] that AXA = A always has solutions, which we denote by $A^-, A^=$ etc. Such matrices are called <u>regular</u>. Similarly, any solution to XAX = X will be denoted by $\hat{A}$, while A^+ will be used to denote any solution common to both. It is often convenient to call these generalized inverses: "inner," "outer" and "reflexive" inverses respectively, or simply 1,2 and 1-2 inverses for short [1]. As always $\mathcal{R}_{m\times n}$ denotes the set of all m × n matrices over $\mathcal{R}$ with $\mathcal{R}^m = \mathcal{R}_{m\times 1}$, ${}^n\mathcal{R} = \mathcal{R}_{1\times n}$.

The right (left) index, if any, of a square matrix A, is defined and denoted by

$$r(A) = \min_k\{R(A^{k+1}) = R(A^k)\}, \quad \ell(A) = \min_\ell\{RS(A^{\ell+1}) = RS(A^\ell)\}.$$

When both are finite they must be equal [5], and the common value is called the index, i(A), of A. The case i(A) = 0, precisely corresponds to A being

(2-sided) invertible (unit for short). For a matrix A with $i(A) = k$, the Drazin inverse, A^d, is the unique solution to the equations [1],

$$A^\ell XA = A^\ell, \quad \ell \geq k, \quad XAX = X, \quad AX = XA. \tag{1.8}$$

When $i(A) = 0$ or 1, A becomes a group member and A^d becomes the so called <u>group inverse</u> $A^\#$ of A. The existence of $A^\#$ is equivalent to the existence of a commutative 1-inverse $A^\sim$ of A, while the existence of A^d (with $i(A) \geq 1$) is equivalent to the existence of $(A^k)^\#$ for some $k \geq 1$.

Whenever the concepts of convergence or spectra are used, we shall assume our rings to be suitable equipped as for example with $\mathcal{R} = \mathbb{C}$.

For $A \in \mathcal{F}_{n\times n}$, $\mathcal{F}$ an algebraically closed field, a general eigenvalue will be denoted by $\lambda(A)$, while for $\mathcal{F} = \mathbb{C}$, its spectral radius is given by $\rho(A)$. We shall write the characteristic and minimal polynomials as

$$\Delta_A(\lambda) = \prod_{t=1}^{s} (\lambda - \lambda_t)^{n_t}, \quad \psi_A(\lambda) = \prod_{t=1}^{s} (\lambda - \lambda_t)^{m_t} \tag{1.9}$$

respectively, where $\sigma_A(\lambda) = \{\lambda_1,\dots,\lambda_s\}$ is the spectrum of distinct eigenvalues of A, and m_t is the <u>index</u> $i(\lambda_t)$ of λ_t with respect to A. It is well-known [1] that $i(0) = i(A)$, so that $(A - \lambda_t I)^\#$ exists exactly when $m_t = 0$ or 1, that is when $N(A - \lambda_t I) = N(A - \lambda_t I)^2$. Moreover, we shall need the spectral theorem [8, p. 104] which states that if $A \in \mathcal{F}_{n\times n}(\mathbb{C}_{n\times n})$ with minimum polynomial ψ, and for any polynomial (function $f(\lambda)$ for which $f^j(\lambda_t)$ $t = 1,\dots,s$, $j = 0,\dots,m_{t-1}$ are well-defined),

$$f(A) = \sum_{t=1}^{s} \sum_{j=0}^{m_{t-1}} f^{(j)}(\lambda_t) Z_t^{\,j}, \tag{1.10}$$

where the $Z_t^{\,j}$ are the spectral components of A associated with λ_t. It is known that [9] [10],

$$Z_t^{\,j} = (A - \lambda_t I)^j [I - (A - \lambda_t I)(A - \lambda_t I)^d]/j! \tag{1.11}$$

In particular if $\lambda_1 = 1$, we write $Z_1^{\,o} = I - (A - I)(A - I)^d$. This matrix as well as $\mathrm{Ni}\ell(A) = A(I - AA^d)$, can in fact be defined whenever A^d and $(A - I)^d$ exist. Throughout this paper we denote similarity by $\approx$.

2. LEMMATA

Since the concept of a group inverse for a square matrix [1] is closely related to the concept of convergence, we begin by recalling some of its lesser known algebraic properties and extending these in part to <u>unit-regular</u> matrices A, for which there exist <u>unit</u> inner inverses A^{-}. Square matrices over a skew-field are examples of this.

<u>Lemma 2.1</u> Let $\mathcal{R}$ be a ring with unity 1 and let $A \in \mathcal{R}_{n\times n}$. The following are equivalent:

1. $A^{\#}$ exists. (2.1)
2. $AXA = A$, $AX = XA$ has a solution $X = A^{\sim}$.
3. $R(A^2) = R(A)$, $RS(A^2) = RS(A)$.
4. $\mathcal{R}^n = R(A) \oplus N(A)$.

In which case

5. $W = A^{\#} + I - AA^{\#}$ is a unit, $W^{-1} = A + I - AA^{\#}$, $AWA = A$ and $AW = WA = AA^{\#}$.

For a proof we refer to [7] [17] [12] [16]. It should thus be noted, that if $A^{\#}$ exists, then A is unit regular.

<u>Lemma 2.2</u> Let $\mathcal{R}$ be a regular ring with unity and let $A, B \in \mathcal{R}_{m\times m}$.

(α) Suppose $R(A) = R(B)$. (2.2)

(i) If $P = C_o + BC_1 + \ldots + B^k C_k$ and $C_o A = AC_o$, then $PAA^{-} \approx PBB^{-}$.

In particular

(ii) $AA^{-} \approx BB^{-}$, $A^2A^{-} \approx ABB^{-}$

$A^2A^{-} + I - AA^{-} \approx ABB^{-} + I + BB^{-}$ and

$AC = CA \Rightarrow A(A - C)(A - C)^{-} \approx A(A - C)(A - C)^{=}$.

(β) If $RS(A) = RS(B)$ and

(i) If $Q = C_o + C_1B + \ldots + C_kB^k$ with $C_oA = AC_o$, then $A^{-}AQ \approx B^{-}BQ$.

In particular

(ii) $A^-A \approx B^-B$, $A^-A^2 \approx B^-BA$

$A^-A^2 + I - A^-A \approx B^-BA + I - B^-B$ and

$AC = CA \Rightarrow (A - C)^-(A - C)A \approx (A - C)^=(A - C)A.$

(γ) $R(A) = R(B)$ and $RS(A) = RS(B)$ exactly when $AU = B = VA$ for some unit matrices U,V.

Proof (α)(i) Since $AA^-B = B$ and $BB^-A = A$ it follows that $W = I - BB^- + AA^- = (I + BB^- - AA^-)^{-1}$. Next $PBB^-W = PAA^- = (I - BB^-)C_oAA^- + [AA^-C_o + BC_1 + \ldots + B^kC_k]AA^- = (I - BB^-)PAA^- + AA^-PAA^- = WPAA^-$.

α(ii) This follows on selecting A,B and P suitable in part (i). (β) follows by symmetry. For (γ) see [14]. □

Lemma 2.3 If A is a unit regular square matrix, then

(i) $AC = CA \Rightarrow CAA^- \approx A^-AC$ (2.3)

(ii) $\left.\begin{array}{l} A^-XA = Y \\ AYA^= = X \\ \text{(pseudo-similarity)} \end{array}\right\} \Rightarrow \begin{array}{c} X \approx Y \\ \text{(similarity)} \end{array}$

In particular

(iii) $AA^- \approx A^-A$, $A^2A^- \approx A^-A^2$, and $A^2A^- + I - AA^- \approx A^-A^2 + I - A^-A$.

Proof (i) If $AA^=A = A$, where $A^=$ is a unit, then $A^=(CAA^=)(A^=)^{-1} = A^=CA = A^=C$, and hence by Lemma (2.2), $CAA^- \approx CAA^= \approx A^=AC \approx A^-AC$.

(ii) See [16] for a proof.

(iii) Combine Lemma 2.2 and 2.3. □

Corollary 2.4 If A and B are unit regular $n \times n$ matrices, then

(i) $R(A) = R(B) \Rightarrow AW_1 = B$ for some unit W_1 (2.4)

(ii) $RS(A) = RS(B) \Rightarrow W_2A = B$ for some unit W_2

Proof (i) Suppose that $A^=$ and $B^=$ are unit 1-inverses of A and B and let $W = I - AA^= + BB^=$. Then W is a unit, $WB = B$ and $AA^=W = BB^= = WBB^=$. This means that $AA^=W(B^=)^{-1} = WB = B$, as desired. (ii) Use symmetry. □

Corollary 2.5 Let $R(A) = R(B)$, and $RS(A) = RS(B)$. If A is unit regular, B is regular, and $AC = CA$, then,

$$CB\bar{\bar{B}} \approx CB\bar{B} \approx CA\bar{A} \approx \bar{A}AC \approx \bar{B}BC \approx \bar{\bar{B}}BC. \tag{2.5}$$

Corollary 2.6 If $A^{\#}$ exists, and if $A\tilde{A} = \tilde{A}A$, $A\tilde{A}A = A$, then

(i) Lemma 2.3 holds. (2.6)

(ii) $R(A) = R(B) \Rightarrow \tilde{A}AA^{-} \approx \tilde{A}BB^{-} \approx A^{\#}AA^{\#} = A$, and

$(I - \tilde{A})AA^{-} \approx (I - \tilde{A})BB^{-} \approx (I - A^{\#})AA^{\#}$.

(iii) $RS(A) = RS(B) \Rightarrow A^{-}A\tilde{A} \approx B^{-}B\tilde{A} \approx A^{\#}$ and

$A^{-}A(I - \tilde{A}) \approx B^{-}B(I - \tilde{A}) \approx A^{\#}A(I - A^{\#})$.

We remark that actually $AA^{-} \approx A^{-}A$ is equivalent to (2.3)-ii or to A being unit regular but that this does not imply that $A^{\#}$ exists [16].

Lemma 2.7 If $\mathcal{R}$ is a ring and $A \in \mathcal{R}_{n\times n}$, and if $A^{k+1}X = A^{k}$, $A^{\ell} = YA^{\ell+1}$, then A^{d} exists and is given by

$$A^{d} = A^{k}X^{k+1} = Y^{\ell+1}A^{\ell} = A^{d}AX = YAA^{d} = YA^{k}X^{k} = Y^{\ell}A^{\ell}X. \tag{2.7}$$

In addition $AA^{d} = A^{k}X^{k} = Y^{\ell}A^{\ell}$.

Proof See [5].

The following identities will be used later in our study of the convergence of the iterations $S_n^{\ k}$ and $T_n^{\ k}$.

Lemma 2.8 Let $\mathcal{R}$ be a regular ring with unity and let $A \in \mathcal{R}_{m\times n}$, $X \in \mathcal{R}_{n\times m}$. Then for $r \geq 1$,

(i) $[(I - XA)XX^{-}]^{r} = (I - XA)^{r}XX^{-} = XX^{-}(I - XA)^{r}XX^{-}$ (2.8)

$= [XX^{-}(I - XA)XX^{-}]^{r}$.

(ii) $[X^{-}X(I - AX)]^{r} = X^{-}X(I - AX)^{r} = X^{-}X(I - AX)^{r}X^{-}X$

$= [X^{-}X(I - AX)X^{-}X]^{r}$.

(iii) $[XX^{-}(I - XA)]^{r} = (I - XA)^{r-1}XX^{-}(I - XA)$

(iv) $[(I - AX)X^{-}X]^{r} = (I - AX)X^{-}X(I - AX)^{r-1}$

(v) $XX^{-}(I - XA)^{r} = (I - XA)^{r} - (I - XX^{-})$

(vi) $(I - AX)^{r}X^{-}X = (I - AX)^{r} - (I - X^{-}X)$.

Proof Induction. □

Lemma 2.9 If $(a)_o = 1$ and $(a)_t = a(a + 1) \ldots (a + t - 1)$, then

$$\text{(i)} \quad \sum_{t=m}^{n} \frac{(a)_t}{t!} = \begin{cases} \dfrac{(a+1)_n}{n!} - \dfrac{(a+1)_{m-1}}{(m-1)!} & \text{if } m \geq 1 \\[2ex] \dfrac{(a+1)_n}{n!} & \text{if } m = 0. \end{cases} \tag{2.9}$$

$$\text{(ii)} \quad \sum_{r=m}^{n} \binom{r}{s} = \binom{n+1}{s+1} - \binom{m}{s+1}$$

$$\text{(iii)} \quad \sum_{r=k}^{n-1} \binom{n-r}{n}\binom{r}{s} = \frac{1}{n}\binom{n+1}{s+2} - \binom{k}{s+1} + \frac{1}{n}\binom{k}{s+1}\left\{\frac{(s+1)k-1}{s+2}\right\}, \quad 0 \leq s \leq k-1$$

Proof (i) Use induction. (ii) Observe that $(s + t)! + s!(s + 1)_t$ and that $\binom{r}{s} = (s + 1)_{r-s}/(r - s)$. (iii) The left hand side becomes $\frac{n+1}{n}\sum_{r=k}^{n-1}\binom{r}{s} - \frac{1}{n}\sum_{r=k}^{n-1}(r+1)\binom{r}{s} = \frac{n+1}{n}\sum_{r=k}^{n-1}\binom{r}{s} - \frac{s+1}{n}\sum_{u=k+1}^{n}\binom{u}{s+1}$

$= \frac{n+1}{n}\left[\binom{r}{s+1} - \binom{k}{s+1}\right] - \frac{s+1}{n}\left[\binom{n+1}{s+2} - \binom{k+1}{s+2}\right]$, which reduces to the desired result. □

3. THE CONVERGENCE BEHAVIOR OF C_n^k, D_n^k

In this section we shall investigate the algebraic, the spectral and the asympotic properties of the Cesaro sum $C_n^k(A) = \sum_{r=0}^{n-1} A^r/n^k$ and the differentiated Cesaro sum $D_n^k(A) = \sum_{r=0}^{n-1} (n - r)A^r/n^{k+1}$, in relation to the matrices A^n/n^k. These properties are useful in the theory of Markov chains [18], [19], and iteration theory [22], and will form the basis to our investigation of the Neumann iterations S_n^k, T_n^k. One of the key steps in this study will be a close analysis of the properties of the matrices $A_c = A(A - C)(A - C)^-$ with $AC = CA$, which appear repeatedly throughout this investigation. For example see Theorems 3.10 and 3.11.

We shall begin with deriving, under general conditions, the algebraic identities relating these matrices, which allow us to <u>compare</u> their convergence behavior and even compute their limit for certain special cases. After this we specialize to complex matrices and use a spectral analysis to derive the <u>actual</u> convergence conditions in terms of eigenvalues. First a preliminary result dealing with A_c.

<u>Theorem 3.1</u> Let $\mathcal{R}$ be a regular ring with unity and let $A, (A - C)^d \in \mathcal{R}_{m\times m}$. Also let

$$A_c = A(A - C)(A - C)^-, \quad A_c' = A(A - C)(A - C)^=,$$

$$B_c = (A - C)^-(A - C)A, \quad B_c' = (A - C)^=(A - C)A, \quad E_c = A(A - C)(A - C)^d$$

If $AC = CA$, then the following hold for $k \geq 1$,

(i) $A_c^k = A^k(A - C)(A - C)^-, B_c^k = (A - C)^-(A - C)A^k$ and (3.1)

$E_c^k = A^k(A - C)(A - C)^d$

(ii) $A_cA_c' = A_c'^2, B_cB_c' = B_c^2$

(iii) $A_c^k(A - C)(A - C)^= = (A_c')^k \quad (A - C)^=(A - C)B_c^k = (B_c')^k$

(iv) If $(A - C)$ is unit regular, then $A_c' \approx A_c \approx B_c \approx B_c'$

(v) If $\mathcal{R}$ is commutative, then $B_c^T = A^T(A^T - C^T)(A^T - C^T)^-$ for some $(A^T - C^T)^-$.

(vi) $(I - A_c + A_c')^{-1} = (I + A_c - A_c')$

$(I - B_c + B_c')^{-1} = (I + B_c - B_c')$.

If in addition C is a central unit, then for $r \geq 1$

(vii) $(A_c - C)^r = (-C)^r[X^{r+1}X^- + I - XX^-]$

$A_c^r = C^r[(I - X^2X^-)^r - (I - XX^-)]$

$(B_c - C)^r = (-C)^r[X^-X^{r+1} + I - X^-X]$

$B_c^r = C^r[(I - X^-X^2)^r - (I - X^-X)]$

$$(E_c - C)^r = (-C)^r[X^{r+1}X^d + I - XX^d]$$

$$E_c{}^r = C^r[(I - X^2X^d)^r - (I - XX^d)], \text{ where } X = I - AC^{-1}.$$

(viii) If $i(A - C) = 1$, there exist equal A_c, B_c, E_c namely

$A(A - C)(A - C)^{\#}$.

(ix) $i(A - C) = \ell \Leftrightarrow i(A_c - C) = \ell - 1, \quad \ell \geq 1$

(x) $i(E_c - C) = 0$.

Proof (i) Use induction. (ii) Obvious. (iii) Follows from (i). (iv) This follows from Corollary 2.5. (v) For matrices over a commutative regular ring $\{M^-\}^T = \{M^T\}^-$. (vi) This follows from (ii). (vii) $A_c - C = (A - C)^2(A - C)^- - C[I - (A - C)(A - C)^-]$
$= (-C)[(I - AC^{-1})^2(I - AC^{-1})^- + I - (I - AC^{-1})(I - AC^{-1})^-]$
$= (-C)[X^2X^- + I - XX^-]$, where $X = I - AC^{-1}$.

Hence by (4.2)-(ii) the result follows. The next lower two identities follow by symmetry. Next, $A_c = (A - C)^2(A - C)^- + C(A - C)(A - C)^-$
$= C[I - (I - AC^{-1})^2(I - AC^{-1})^- - I + (I - AC^{-1})(I - AC^{-1})^-]$
$= C[(I - X^2X^-) - (I - XX^-)]$, from which the fourth result follows by induction. The remaining two identities again follow by symmetry, once we observe that if $(XC)^d$ and C^{-1} exist, then $XC = CX$ ensures that X^d exists and equals $C(XC)^d$. (vii) Obvious. (ix) This follows immediately from Theorem 4.6 and the fact that $i(X) = i(I - AC^{-1}) = i(A - C)$. (x) Note that $(X^2X^d + I - XX^d)^{-1} = X^d + I - XX^d$. □

It is well-known that the geometric progression is the key to all iteration theory, the basic result being the following telescoping identity for square matrices A and C:

$$\sum_{r=0}^{n-1} A^r(A - C) \equiv A^n - I + \sum_{r=0}^{n-1} A^r(I - C). \tag{3.2}$$

If $(A - C)$ is regular, then this implies that

$$(A^n - I)(A - C)^-(A - C) \equiv A^n - I + \sum_{r=0}^{n-1} A^r(I - C)[I - (A - C)^-(A - C)] \tag{3.3}$$

and moreover by (3.1)-i, if $AC = CA$,

$$\sum_{r=0}^{n-1} A^r(A - C)(A - C)^- \equiv (A^n - I)(A - C)^- + (I - C) \sum_{r=0}^{n-1} A^r(A - C)^- \quad (3.4)$$

$$\equiv \sum_{r=0}^{n-1} [A(A - C)(A - C)^-]^r + (A - C)(A - C)^- - I.$$

Most useful are the corresponding results with B = I - A.

$$\sum_{r=0}^{n-1} (I - B)^r B \equiv I - (I - B)^n, \quad \text{and} \quad (3.5)$$

$$[I - (I - B)^n]B^- \equiv \sum_{r=0}^{n-1} (I - B)^r BB^- = \sum_{r=0}^{n-1} [(I - B)BB^-]^r + BB^- - I. \quad (3.6)$$

Formally <u>differentiating</u> the telescoping identity with respect to A we obtain

$$\sum_{r=0}^{n-1} rA^r(A - C) \equiv nA^n + \sum_{r=0}^{n-1} rA^r(I - C) - A \sum_{r=0}^{n-1} A^r. \quad (3.7)$$

and hence

$$\sum_{r=0}^{n-1} (n - r)A^r(A - C) \equiv -nI + A \sum_{r=0}^{n-1} A^r + \sum_{r=0}^{n-1} (n - r)A^r(I - C). \quad (3.8)$$

This implies that for regular (A - C),

$$\sum_{r=0}^{n-1} (n - r)A^r(A - C) \equiv -n(A - C)^-(A - C) + A \sum_{r=0}^{n-1} A^r(A - C)^-(A - C)$$

$$+ \sum_{r=0}^{n-1} (n - r)A^r(I - C)(A - C)^-(A - C) \quad (3.9)$$

and for AC = CA,

$$\sum_{r=0}^{n-1} (n - r)A^r(A - C)(A - C)^- \equiv -n(A - C)^- + A \sum_{r=0}^{n-1} A^r(A - C)^-$$

$$+ (I - C) \sum_{r=0}^{n-1} (n - r)A^r(A - C)^- \equiv \sum_{r=0}^{n-1} (n - r)[A(A - C)(A - C)^-]^r$$

$$+ n(A - C)(A - C)^- - nI. \quad (3.10)$$

In particular, if B = I - A, then

$$\sum_{r=0}^{n-1} (n - r)(I - B)^r B \equiv nI - (I - B) \sum_{r=0}^{n-1} (I - B)^r \quad \text{and} \quad (3.11)$$

$$\sum_{r=0}^{n-1} (n - r)(I - B)^r BB^- \equiv -nB^- - (I - B) \sum_{r=0}^{n-1} (I - B)^r B^-, \tag{3.12}$$

which we shall use later on.

The left hand identities corresponding to (3.2)-(3.12) follow trivially. We remark, that the above identities are really equivalent to the corresponding ones with $C = I$, because the terms containing C in (3.2) and (3.8) identically vanish when collected.

We continue now by observing that if $B_n = A^n$ and $C_n = \sum_{r=0}^{n-1} [A(A - I)(A - I)^-]^r = C_n^{\,o}(A_1)$, then (3.2) and (3.4) yield

$$B_n = C_n(A - I) + I \qquad C_n = B_n(A - I)^- + I - A(A - I)^-. \tag{3.12a}$$

This is a special case of the equivalence relation "being algebraically related," which we define for sequences in arbitrary rings by

$$b_n \sim c_n \Leftrightarrow \begin{cases} b_n = c_n p + q \\ c_n = b_n r + s, \text{ for some } p, q, r, s. \end{cases} \tag{3.13}$$

For complex matrices and matrices over a Hausdorff (topological) skew-field in general, [4, p. 174] the above equivalence relation is too restrictive to be very useful. We shall frequently see cause to use the weaker equivalence relation "weakly algebraically related," w.a.r. for short:

$$B_n \sim C_n \Leftrightarrow \begin{cases} B_n = C_n P_n + Q_n & P_n, Q_n \text{ convergent} \\ C_n = B_n R_n + S_n & R_n, S_n \text{ convergent} \end{cases} \tag{3.14}$$

For example, using (3.12a) it follows immediately that for all $k \geq 0$,

$$A^n/n^k = C_n^{\,k}(A_1) \cdot (A - I) + I/n^k \tag{3.15a}$$

$$C_n^{\,k}(A_1) = A^n/n^k \cdot (A - I)^- + [I - A(A - I)^-]/n^k, \tag{3.15b}$$

while from (3.8) and (3.10) respectively, we have for $k \geq 0$,

$$AC_n^{\,k}(A)/n = D_n^{\,k}(A_1) \cdot (A - I) + I/n^k \tag{3.16a}$$

$$D_n^{\,k}(A_1) = A[C_n^{\,k}(A)/n](A - I)^- + [I - A(A - I)^-]/n^k. \tag{3.16b}$$

Because we also have the identities:

$$AC_n^k(A)/n \equiv (1 + 1/n)^{k+1}C_{n+1}^{k+1}(A) - I/n^{k+1} \tag{3.17a}$$

$$AC_n^k(A_1)/n \equiv (1 + 1/n)^{k+1}C_{n+1}^{k+1}(A_1) + [A - I - A_1]/n^{k+1}, \tag{3.17b}$$

we see that A^n/n^k converges if and only if $C_n^k(A_1)$ converges, while $D_n^k(A_1)$ converges exactly when $AC_n^k(A)/n$ converges, or exactly when $C_{n+1}^{k+1}(A)$ or $C_n^{k+1}(A)$ converges. We note that $D_n^{-1}(A_1)$ will <u>not</u> converge in general as seen from (3.10).

It is easily seen that if $A^n/n^k \to L$ and $C_n^k(A_1) \to M$, then

$$\begin{aligned} L &= M(A - I) + I \cdot \delta_{k,o} = L(A - I)^-(A - I) + [I - (A - I)^-(A - I)]\delta_{k,o} \\ M &= L(A - I)^- + [I - A(A - I)^-]\delta_{k,o} = M(A - I)^-(A - I) \\ &\qquad + [I - (A - I)^-(A - I)]\delta_{k,o}. \end{aligned} \tag{3.18}$$

In particular, for $k \geq 1$, $L = 0 \Leftrightarrow M = 0$. Similar conclusions can be drawn when $AC_n^k(A)/n \to L$ and $D_n^k(A) \to M$.

One of our aims is to establish that if $k \geq i(A - I)$, then A^n/n^k, $C_n^k(A)$ and $C_n^k(A_1)$ are <u>all</u> weakly related. Most of this is easy, since for all $k \geq 0$,

$$C_n^k(A_1) = C_n^k(A) \cdot (A - I)(A - I)^- + [I - (A - I)(A - I)^-]/n^k \tag{3.19}$$

and

$$C_n^k(A_1)(A - I) = C_n^k(A) \cdot (A - I) = (A^n - I)/n^k. \tag{3.20}$$

Consequently, rewriting this yields:

$$A^n/n^k = C_n^k(A) \cdot (A - I) + I/n^k. \tag{3.21}$$

The last remaining link expressing $C_n^k(A)$ in terms of A^n/n^k will be given as a corollary to the next theorem. Next we shall establish that $D_n^k(A)$, A^{n+1}/n^{k+1} and $AC_n^k(A_1)/n$ are all weakly algebraically related. A first step towards this is the following:

$$D_n^k(A_1) \equiv D_n^k(A) \cdot (A - I)(A - I)^- + [I - (A - I)(A - I)^-]/n^k \tag{3.19a}$$

$$D_n^k(A_1) \cdot (A - I) = D_n^k(A)(A - I) = AC_n^k(A)/n - I/n^k, \tag{3.20a}$$

which may be multiplied by (A - I) to give

$$AC_n^k(A) \cdot (A - I)/n = A(A^n - I)/n^{k+1} = D_n^k(A) \cdot (A - I)^2/n^k + (A - I)/n^k. \tag{3.21a}$$

We are now in a position to state our main conclusion based on the above identities for $m \times m$ matrices over a Hausdorff skew field $\mathcal{D}$.

Theorem 3.2 Let $A \in \mathcal{D}_{m\times m}$, $A \neq I$ and let $\ell = i(A - I)$. Then the following implications hold for $k \geq 0$.

(α) (i) $C_n^k(A)$ converges $\Rightarrow$ (ii) $C_n^k(A_1)$ converges $\Leftrightarrow$ (iii) A^n/n^k converges. (3.22)

If $k \geq \ell$, these are all equivalent and the three expressions are all weakly algebraically related.

(β) The following are equivalent

(i) $A^n/n^k \to 0$ (ii) $A^n(A - I)^-/n^k \to 0$ for all $(A - I)^-$

(iii) For all $(A - I)^-$, $C_n^k(A_1) \to \begin{cases} I - A(A - I)^- & \text{if } k = 0 \\ 0 & \text{if } k > 0. \end{cases}$

When $k > \ell$ these are equivalent to

(iv) $C_n^k(A) \to \begin{cases} (I - A)^{-1} & \text{if } k = \ell = 0 \\ 0 & \text{if } k > \ell = 0 \\ \frac{1}{\ell} Z_1^{\ell-1} & \text{if } k = \ell \geq 1 \\ 0 & \text{if } k > \ell \geq 1 \end{cases}$

(γ) (i) $D_n^k(A)$ converges $\Rightarrow$ (ii) $D_n^k(A_1)$ converges $\Leftrightarrow$ (iii) $AC_n^k(A)/n$ converges $\Leftrightarrow$ (iv) $C_{n+1}^{k+1}(A)$ converges $\Rightarrow$ (v) $AC_n^k(A_1)/n$ converges $\Leftrightarrow$ (vi) $C_{n+1}^{k+1}(A_1)$ converges $\Leftrightarrow$ (vii) A^{n+1}/n^{k+1} converges.

When $k \geq \ell$ these are all equivalent and the expressions are weakly algebraically related.

(δ)

$$\text{(i)} \quad D_n^{\,k}(A) \to \begin{cases} (I - A)^{-1} & \text{if } k = 0 \\ \dfrac{1}{\ell(\ell + 1)} Z_1^{\,\ell-1} & \text{if } k \geq 1 \end{cases}$$

$$\Rightarrow \text{(ii)} \quad D_n^{\,k}(A_1) \to \begin{cases} (I - A)^{-1} & \text{if } k = 0 \\ 0 & \text{if } k \geq 1 \end{cases}$$

$$\Leftrightarrow \text{(iii)} \quad AC_n^{\,k}(A)/n \to 0 \Rightarrow \text{(iv)} \quad AC_n^{\,k}(A_1)/n \to 0 = \text{(v)} \quad A^{n+1}/n^{k+1} \to 0.$$

If $k \geq \ell$, these are all equivalent.

<u>Proof</u> Let us first give the easy parts.
(α) (i) ⇒ (ii) This follows from (3.19). (ii) ⇔ (iii) Clear by (3.15).
(β) (i) ⇒ (ii) Obvious. (ii) ⇒ (i) For $k \geq 1$ this follows from the identity $(A^n - I)(A - I)^-(A - I)/n^k \equiv (A^n - I)/n^k$, while for $k = 0$, we note that if $(A - I)^-$ is an inner inverse of $A - I$, then so is $(A - I)^- + I - (A - I)^-(A - I)$.

Hence $A^n(A - I)^-(A - I) \to 0$ as well as $A^n[I - (A - I)^-(A - I)] \to 0$, ensuring that $A^n \to 0$. (ii) ⇔ (iii) Since $A^n(A - I)^- \to 0 \Leftrightarrow (A^n - I)(A - I)^- \to (I - A)^-$, this follows from (3.4).
(γ) (i) ⇒ (ii) Clear by (3.19a). (ii) ⇔ (iii) Simply use (3.16).
(iii) ⇔ (iv) Apply (3.17a). (iii) ⇒ (v) This follows from (3.19).
(v) ⇔ (vi) Use (3.17b). (vi) ⇔ (vii) Follows from (3.15).
(δ) (i) ⇒ (ii) If $(A - I)^{-1}$ exists, then $A_1 = A$ so that the result is clear for $k = 0$.

For $k \geq 1$, the result follows from (3.19a) and the fact that $(A - I)Z_1^{\,\ell-1} = 0$ if $\ell = i(A - I)$.

Note that for $k \geq \ell$, (i) can be written more conveniently as

$$D_n^{\,k}(A) \to \begin{cases} (I - A)^{-1} & \text{if } k = \ell = 0 \\ 0 & \text{if } k > \ell = 0 \\ \dfrac{1}{\ell(\ell + 1)} Z_1^{\,\ell+1} & \text{if } k = \ell \geq 1 \\ 0 & \text{if } k > \ell \geq 1, \end{cases} \tag{3.22a}$$

which will be used to facilitate the computation of $\lim T_n^k$.

(ii) $\Leftrightarrow$ (iii) The necessity is clear by (3.16a). For sufficiency we note that if $AC_n^o(A)/n \to 0$ then by (3.16a) $D_n^o(A_1)(I - A) \to I$, which not only implies that $(I - A)^{-1}$ exists, but also that $D_n^o(A_1) \to (I - A)^{-1}$. For $k \geq 1$, the sufficiency follows from (3.16b). (iii) $\Rightarrow$ (iv) Clear by (3.19). (iv) $\Leftrightarrow$ (v) This follows from (3.15) on multiplication by A and division by n. Let us now turn to the missing links in the proof. The essential part is contained in the following algebraic results.

Lemma 3.3 Let $\mathcal{R}$ be a ring with unity, and $I \neq A \in \mathcal{R}_{m\times m}$ such that $i(A - I) = \ell$. Then

$$\text{(i)} \quad \text{If } \ell = 0, \sum_{r=0}^{n-1} A^r = (A^n - I)(A - I)^{-1} \tag{3.23}$$

$$\text{(ii)} \quad \text{If } \ell = 1, \sum_{r=0}^{n-1} A^r = (A^n - I)(A - I)^{\#} + n[I - (A - I)(A - I)^{\#}]$$

$$\begin{aligned}\text{(iii)} \quad \text{If } \ell \geq 2, \sum_{r=0}^{n-1} A^r &= (A^n - I)(A - I)^d \\ &+ \sum_{r=0}^{\ell-1} A^r[I - (A - I)(A - I)^d] \\ &+ \sum_{r=\ell}^{n-1} \sum_{s=0}^{\ell-2} \binom{r}{s} (A - I)^s[I - (A - I)(A - I)^d] \\ &+ \left\{\binom{n}{\ell} - 1\right\}(A - I)^{\ell-1}[I - (A - I)(A - I)^d].\end{aligned}$$

Proof It suffices to prove (iii), if we assume the usual convention concerning empty sums. Setting C = I in (3.2) and multiplying by $(A - I)^d$, we obtain first of all

$$(A^n - I)(A - I)^d \equiv \sum_{r=0}^{n-1} A^r(A - I)(A - I)^d \quad \text{and hence}$$

$$\begin{aligned}\sum_{r=0}^{n-1} A^r = \sum_{r=0}^{n-1} A^r[(A - I)(A - I)^d + Z_1^o] &= (A^n - I)(A - I)^d \\ &+ \left\{\sum_{r=0}^{\ell-1} + \sum_{r=\ell}^{n-1}\right\} A^r Z_1^o,\end{aligned}$$

where, as in (1.10) $Z_1^o = I - (A - I)(A - I)^d$.

Writing $A = (A - I) + I$, and expanding, the last sum may be written as

$$\sum_{r=\ell}^{n-1} \sum_{s=0}^{\ell-1} \binom{r}{s} (A - I)^s Z_1^o$$

because $(A - I)^s Z_1^o \equiv 0$ for $s \geq \ell = i(A - I)$. Extracting the term with $s = \ell - 1$, $\ell \geq 1$, yields: $\sum_{r=\ell}^{n-1} \binom{r}{\ell - 1} (A - I)^{\ell-1} Z_1^o$, in which the sum reduces with aid of Lemma (2.10)ii to $\binom{n}{\ell} - 1$, completing the proof. □

For matrices over a skew-Hausdorff field, we have an immediate Corollary that for all $k \geq 0$,

$$C_n^k(A) \equiv A^n(A - I)^d/n^k - (A - I)^d/n^k + \sum_{r=0}^{\ell-1} A^r Z_1^o/n^k \tag{3.24}$$
$$+ \sum_{r=\ell}^{n-1} \sum_{s=0}^{\ell-2} \binom{r}{s}(A - I)^s Z_1^o/n^k + [\binom{n}{\ell} - 1](A - I)^{\ell-1} Z_1^o/n^k.$$

The second and third terms both converge (to zero if $k \geq 1$) as $n \to \infty$. Now since $\binom{r}{s} = 0\left(\frac{r^s}{s!}\right)$ and $\sum_{r=\ell}^{n-1} r^s = 0(n^{s+1})$ with $s \leq \ell - 2$, it follows that the fourth term in (3.24) is of order $0(n^{-k-1+\ell})$. In the last term we further have

$$\frac{1}{n^k} [\binom{n}{\ell} - 1] = \frac{n^{\ell-k}}{\ell!} \prod_{t=0}^{\ell-1} (1 - t/n) - \frac{1}{n^k},$$

which converges for $k > \ell$ to zero, and for $k = \ell$ to $\frac{1}{\ell!}$.

Hence we have established that A^n/n^k and $C_n^k(A)$ are weakly algebraically related and thus have the same convergence behavior for all $k \geq i(A - I)$. Moreover, for these values of k

$$C_n^k(A) - A^n(A - I)^d/n^k \to (I - A)^d \delta_{k,o} + \sum_{r=0}^{\ell-1} A^r Z_1^o \delta_{k,o} + \frac{1}{\ell} Z_1^{\ell-1} \delta_{k,\ell}, \tag{3.25}$$

where $Z_1^{-1} = 0$. The second term on the right also vanishes since $k \geq \ell$, $k = 0 \Rightarrow \ell = 0$. We have thus shown that in (3.22) (αiii) ⇒ (αi) and (βi) ⇒ (βiv). Lastly (3.21) shows that (βiv) ⇒ (βi) when $k \geq \ell$, completing these parts.

Before continuing with part (γ), we should remark that for $k < \ell$ one cannot in general expect similar convergence behavior for A^n/n^k and $C_n^{\ k}(A)$, as seen from (3.23)ii with $k = 0$, $\ell = 1$. It should further be noted here, that (3.23)ii may be obtained much more directly on summing the identity:

$$A^r - [I - (A - I)(A - I)^{\#}] \equiv [A(A - I)(A - I)^{\#}]^r, \quad r \geq 1, \tag{3.26}$$

of (3.1)-vii, which when combined with (3.4) yields the desired result.

Let us now turn to the convergence behavior of $D_n^{\ k}(A)$ and complete the proof of (γ) and (δ).

<u>Lemma 3.4</u> Let $\mathcal{R}$ be a ring with unity and let $A \in \mathcal{R}_{m\times m}$ such that $i(A - I) = \ell$. Then

(i) If $\ell = 0$, $\sum_{r=0}^{n-1} (n - r)A^r \equiv n(I - A)^{-1} + A(A^n - I)(A - I)^{-2}$

(ii) If $\ell = 1$, $\sum_{r=0}^{n-1} (n - r)A^r \equiv n(I - A)^{\#} + A(A^n - I)[(A - I)^{\#}]^2$

$$+ \frac{1}{2} n(n + 1)[I - (A - I)(A - I)^{\#}]$$

(iii) If $\ell \geq 2$, $\sum_{r=0}^{n-1} (n - r)A^r \equiv n(I - A)^d + A(A \ \ - I)[(A - I)^d]^2$

$$+ \sum_{r=0}^{\ell-1} (n - r)A^r Z_1^{\ o} + \sum_{s=0}^{\ell-2} (A - I)^s Z_1^{\ o} \left[\binom{n + 1}{s + 2} - n\binom{\ell}{s + 1}\right.$$

$$\left. + \binom{\ell}{s + 1}\left\{\frac{\ell(s + 1) - 1}{s + 2}\right\}\right] + (A - I)^{\ell-1} Z_1^{\ o}\left[\binom{n + 1}{\ell + 1} - n + \ell - 1\right]. \tag{3.27}$$

<u>Proof</u> For general ℓ we begin by multiplying (3.8) with $C = I$, by $(A - I)^d$ and obtain $\sum_{r=0}^{n-1} (n - r)A^r(A - I)(A - I)^d = \left(-nI + A \sum_{r=0}^{n-1} A^r\right)(A - I)^d$ $= [-nI + A(A^n - I)(A - I)^d](A - I)^d$, where we used (3.2) and the 2-inverse property of $(\cdot)^d$. Next, consider

$$\sum_{r=0}^{n-1} (n - r)A^r = \sum_{r=0}^{n-1} (n - r)A^r[(A - I)(A - I)^d + Z_1^{\ o}] = -n(A - I)^d$$

$$+ A(A^n - I)[(A - I)^d]^2 + \left\{\sum_{r=0}^{\ell-1} + \sum_{r=\ell}^{n-1}\right\} (n - r)A^r Z_1^{\ o}.$$

Writing $A = (A - I) + I$ and expanding we may rewrite the last sum as

$$\sum_{r=\ell}^{n-1} (n - r) \sum_{s=0}^{\ell-1} \binom{r}{s}(A - I)^s Z_1^{\,o},$$

since again the terms with $s = \ell,\ldots,r$, vanish. Interchanging summations now yields

$$\sum_{s=0}^{\ell} (A - I)^s Z_1^{\,o} \sum_{r=\ell}^{n-1} (n - r)\binom{r}{s}$$

in which the second term may be computed with aid of Lemma (2.10)ii. Extracting the term with $s = \ell - 1$ completes the proof. □

Again for matrices over a Hausdorff-skew-field, we obtain:

$$\begin{aligned} D_n^{\,k}(A) = {} & \frac{A^{n+1}}{n^{k+1}} [(A - I)^d]^2 - A[(A - I)^d]^2/n^{k+1} + (I - A)^d/n^k \\ & + \sum_{r=0}^{\ell-1} \frac{n - r}{n} A^r Z_1^{\,o}/n^k + \sum_{s=0}^{\ell-2} (A - I)^s Z_1^{\,o}\left[\frac{1}{n}\binom{n + 1}{s + 2} - \binom{\ell}{s + 1}\right. \\ & \left. + \frac{1}{n}\binom{\ell}{s + 1}\left\{\frac{(s + 1)\ell - 1}{s + 2}\right\}\right]/n^k + (A - I)^{\ell-1} Z_1^{\,o}\left[\frac{1}{n}\binom{n + 1}{\ell + 1}\right. \\ & \left. - 1 + (\ell - 1)/n\right]/n^k, \end{aligned} \tag{3.28}$$

in which the second, third and fourth terms obviously all converge. Now consider

$$\begin{aligned} \binom{n + 1}{s + 1} &= \frac{n + 1}{(s + 1)!} n^s \prod_{t=0}^{s-1} (1 - t/n) = \frac{(n + 1)}{(s + 1)!} n^s \sum_{r=0}^{s-1} \frac{(-1)^r}{n^r} \sigma_r^{(s-1)} \\ &= \frac{n^{s+1}}{(s + 1)!} \sum_{r=0}^{s-1} \frac{(-1)^r}{n^r}\left[\sigma_r^{(s-1)} - \sigma_{r-1}^{(s-1)}\right], \end{aligned} \tag{3.29}$$

where $\sigma_r^{(s-1)}$ is the rth symmetric function of the roots $\{0,1,2,\ldots,s - 1\}$, $\sigma_o^{(s-1)} = 1$ and $\sigma_{-1}^{(s-1)} = \sigma_s^{(s-1)} = 0$. (The $\sigma_r^{(s)}$ are closely related to the Stirling numbers of the first kind.) Hence

$$\frac{1}{n^{k+1}} \binom{n + 1}{s + 2} \sim \frac{n^{s+1-k}}{(s + 2)!}, \tag{3.30}$$

which converges for $k \geq s + 1$. In the fifth term $s + 1 \leq \ell - 1$ and hence for $k + 1$, this term vanishes with n. In the last term we obtain:

$$\frac{1}{n^{k+1}}\binom{n+1}{\ell+1} \sim \frac{n^{\ell-k}}{(\ell+1)!} \to \begin{cases} 0 & \text{if } k > \ell \\ \dfrac{1}{(\ell+1)!} & \text{if } k = \ell \\ \text{diverges} & \text{if } k < \ell, \end{cases} \tag{3.31}$$

so that this term yields a limit of $\frac{1}{\ell(\ell+1)} Z_1^{\ell-1}$ if $k = \ell$ and vanishes with n if $k > \ell$. We have thus established, for $k \geq \ell$, that in (3.22), $(\gamma)(vii) \Rightarrow \gamma(i)$, but also that all these terms in (γ) are weakly algebraically related. Lastly, (3.28) shows that for $k \geq \ell$,

$$D_n^k(A) - \frac{A^{n+1}}{n^{k+1}}[(A - I)^d]^2 \to (I - A)^d \delta_{k,o} + \frac{1}{\ell(\ell+1)} Z_1^{\ell-1} \delta_{k,\ell}, \tag{3.32}$$

which shows that $(\delta(v)) \Rightarrow (\delta(i))$ in (3.22), thereby completing the proof of Theorem (3.2). □

We shall now specialize to complex matrices, so that we can use spectral analysis to see under what actual conditions the polynomials $C_n^k(A)$ and $D_n^k(A)$ converge. A principal role is played by the spectrum of the matrix $A_1 = A(A - I)(A - I)^-$, which we shall have to examine first.

Shift lemma 3.7 Suppose $A \in F_{n\times n}$, $\gamma \in \sigma_A$ and let $A_\gamma = A(A - \gamma I)(A - \gamma I)^-$, $B_\gamma = (A - \gamma I)^-(A - \gamma I)A$ and $E_\gamma = A(A - \gamma I)(A - \gamma I)^d$. Then

(i) The lists of elementary divisors of A_γ and B_γ are equal. These in turn equal the list of elementary divisors of A, except that the degree of the elementary divisors corresponding to eigenvalue $\lambda = \gamma$, has been decreased by unity in going from A to A_γ. In particular, (3.33)

$$\Delta_{A_\gamma}(\lambda) = \Delta_{B_\gamma}(\lambda) = \Delta_A(\lambda)\left[\frac{\lambda}{\lambda - \gamma}\right]^{\nu_\gamma}$$

$$\Delta_{E_\gamma}(\lambda) = \Delta_A(\lambda)\left[\frac{\lambda}{\lambda - \gamma}\right]^{n_\gamma}$$

$$\psi_A(\lambda) = \psi_{B_\gamma}(\lambda) = \mathrm{LCM}\left(\lambda, \frac{\psi_A(\lambda)}{\lambda - \gamma}\right) = \begin{cases} \psi_A/(\lambda - \gamma) & \text{if A is singular} \\ \lambda\psi_A/(\lambda - \gamma) & \text{if A is invertible} \end{cases}$$

$$\psi_{E_\gamma}(\lambda) = \mathrm{LCM}\left(\lambda, \frac{\psi_A(\lambda)}{(\lambda - \gamma)^\ell}\right) = \begin{cases} \psi_A(\lambda)/(\lambda - \gamma)^\ell & \text{if A is singular} \\ \lambda\psi_A(\lambda)/(\lambda - \gamma)^\ell & \text{if A is invertible} \end{cases}$$

where n_γ, ν_γ are the algebraic and geometric multiplicities of the eigenvalue γ, and $\ell = i(A - \gamma I) \geq 1$.

(ii) $i(A - \gamma I) = 1 \Leftrightarrow i(A_\gamma - \gamma I) = 0 \Leftrightarrow \Delta_{A_\gamma}(\lambda) = \Delta_A(\lambda)\left(\frac{\lambda}{\lambda - \gamma}\right)^{n_\gamma}$

$\Leftrightarrow \sigma_{A_\gamma} = \{\sigma_A/\gamma\} \cup \{0\}.$

(iii) The results of (3.1) are valid with $C = \gamma I$, $\gamma \neq 0$.

Proof (i) Since $A_\gamma \approx B_\gamma$ it is clear that their characteristic and minimal polynomials are equal. Also, by (3.1)iv, these polynomials are invariant under the choice of $(A - \gamma I)^-$, and so we may take advantage of this by selecting a particularly simple $(A - \gamma I)^-$. Let $i(A - \gamma I) = \ell \geq 1$ and consider the Jordan or Jacobson canonical form of A, (which agree in the γ-Jordan blocks). $J_A = J_\ell(\gamma) \oplus \ldots \oplus J_t(\gamma) \oplus K$, where

$$J_\ell(\gamma) = \begin{bmatrix} \gamma & 1 & & 0 \\ & \ddots & \ddots & \\ & & \ddots & 1 \\ 0 & & & \gamma \end{bmatrix}_{\ell\times\ell}$$

, and $\ell \geq t \geq 1$ and $(K - \gamma I)$ is invertible. Hence

$(J_A - \gamma I) = J_\ell(0) \oplus \ldots \oplus J_t(0) \oplus (K - \gamma I)$. As a particular inner inverse we take

$$(J_A - \gamma I)^- = J_\ell^T(0) \oplus \ldots \oplus J_t^T(0) \oplus (K - \gamma I)^{-1}, \tag{3.34}$$

and hence it now suffices to consider just one Jordan block in $J_A(J_A - \gamma I)(J_A - \gamma I)^-$. Now $J_\ell(\gamma)[J_\ell(\gamma) - \gamma I][J_\ell(\gamma) - \gamma I]^- =$

$$\begin{bmatrix} \gamma & 1 & & 0 \\ & \ddots & \ddots & \\ & & \ddots & 1 \\ 0 & & & \gamma \end{bmatrix} \left[\begin{array}{c|c} I_{\ell-1} & 0 \\ \hline 0 & 0 \end{array}\right] = \left[\begin{array}{c|c} J_{\ell-1}(\gamma) & 0 \\ \hline 0 & 0 \end{array}\right]_{\ell\times\ell}, \tag{3.35}$$

so that in each γ-Jordan block a factor of $\lambda - \gamma$ has been dropped, from which the first and the third results easily follow. A similar consideration shows that the Jordan form of E_γ is obtained from that of A by replacing all the γ-Jordan blocks by zero blocks. Hence the remaining results follow. In particular $i(E_\gamma) = \max(1, i(A))$. We shall make repeated use of this shift part of the lemma.

(ii) Recall that $i(A - \gamma I) = 1 \Leftrightarrow \nu_\gamma = n_\gamma$, which by (i) implies that $\Delta_{A_\gamma} = \Delta_A\left(\frac{\lambda}{\lambda - \gamma}\right)^{n_\gamma}$. Conversely, if Δ_{A_γ} has this form, then $A_\gamma - \gamma I$ is invertible, which in turn implies by (3.1)ix that $i(A - \gamma I) = 1$. Lastly, this form of Δ_{A_γ} clearly implies that $\sigma_{A_\gamma} = \{\sigma_A \backslash \gamma\} \cup \{0\}$, while conversely, if this holds $(A_\gamma - \gamma I)$ is invertible which implies the desired form for Δ_{A_γ}. Since $\sigma_{E_\gamma} = \{\sigma_A \backslash \gamma\} \cup \{0\}$, $E_\gamma - \gamma I$ is always invertible which checks (3.1)x. □

Remarks 1. If for example, $F = ¢$ and $|\lambda(A)| \leq 1$, then we may use the shift lemma to successively shift all eigenvalues from the unit circle to the origin. For example, if $i(A - \gamma I) = \ell$, and $|\gamma| = 1$, then we may iterate the mapping:

$$A \to A_\gamma = A(A - \gamma I)(A - \gamma I)^- \tag{3.36}$$

ℓ times, that is, $A \to A_\gamma \to A_{\gamma\gamma} = A_\gamma(A_\gamma - \gamma I)(A_\gamma - \gamma I)^- \to A_{\gamma..\gamma}$. This reduces the index of γ by unity at each step until it reaches zero. This means that γ is no longer an eigenvalue of $A_{\underbrace{\gamma..\gamma}_{\ell}}$ and has been "replaced" by extra eigenvalues at 0. All the other eigenvalues including multiplicities and indices have reamined unchanged. Repeating this with all other eigenvalues on the unit circle yields eventually a matrix with spectral radius less than unity. We could also use the map $A \to E_\gamma = A(A - \gamma I)(A - \gamma I)^d$ to do this, but Drazin inverses are of course much harder to compute than 1-inverses.

2. The similarity of A_γ and B_γ may also be seen from the fact that $Y^- A_\gamma Y = B_\gamma$, and $Y B_\gamma Y^- = A_\gamma$, where $Y = A - \gamma I$, which shows that A_γ and B_γ are pseudo-similar via the unit regular matrix $A - \gamma I$ and hence by Lemma (2.3)ii are similar.

We are now ready to sharpen Theorem (3.2) under the extra assumptions on our field F.

Theorem 3.8 If $C_n^k(\lambda) = \sum_{r=0}^{n-1} \lambda^r/n^k$, $A \in ¢_{m\times m}$, $A \neq I$, and $\ell = i(A - I)$, then the following are equivalent for all $k \geq 0$.

(i) $C_n^k(A)$ converges. (ii) $|\lambda(A)| \leq 1$, $|\lambda(A)| = 1 \Rightarrow i(\lambda(A)) \leq k$.

(iii) $A^n/n^k \to 0$. (iv) For all $(A - I)^-$, $A^n(A - I)^-/n^k \to 0$.

(v) For all $(A - I)^-$: $C_n^k(A_1) \to \begin{cases} I - A(A - I)^- & \text{if } k = 0 \\ 0 & \text{if } k \geq 1. \end{cases}$

(vi) $|\lambda(A_1)| \leq 1$, $|\lambda(A_1)| = 1 \Rightarrow i(\lambda(A_1)) \leq k$, $i(A_1 - I) \leq k - 1$.

(vii) $C_n^k(E_1)$ converges, and $i(A - I) \leq k$. (viii) $A^n(A - I)^d/n^k \to 0$ and $i(A - I) \leq k$.

In which case $k \geq \ell$ and

$$\text{(ix)} \quad C_n^k(A) \to \begin{cases} (I - A)^{-1} & \text{if } k = \ell = 0 \\ 0 & \text{if } k > \ell = 0 \\ \frac{1}{\ell} Z_1^{\ell-1}(A) & \text{if } k = \ell \geq 1 \\ 0 & \text{if } k > \ell \geq 1. \end{cases} \tag{3.37}$$

Proof (i) $\Leftrightarrow$ (ii) From the spectral theorem (or the Jordan form) we know that

$$C_n^k(A) \text{ converges} \Leftrightarrow [C_n^k(\lambda)]_{\lambda_t}^{(j)} \quad j = 0,\ldots,i(\lambda_t) - 1 \text{ converges.}$$

Now $[C_n^k(\lambda)]^{(j)} = \frac{1}{n^k} \sum_{r=0}^{n-1} r(r - 1) \ldots (r - j + 1)\lambda^{r-j}$, which clearly diverges for $|\lambda| > 1$ and converges for $|\lambda| < 1$ (to zero if $k \geq 1$). If $\ell \geq 1$ and $\lambda = 1$, then

$$[C_n^k(1)]^{(j)} = \frac{1}{n^k} \sum_{t=0}^{n-j-1} \frac{(t + j)!}{t!} = \frac{j!}{n^k} \sum_{t=0}^{n-j-1} \frac{(j + 1)_t}{t!},$$

which by (2.10) reduces to

$$\frac{j!}{n^k} \frac{(j + 2)_{n-j-1}}{(n - j - 1)!} = \frac{1}{j + 1} \frac{n(n - 1) \ldots (n - j)}{n^k}. \tag{3.38}$$

This converges exactly when $j \leq k - 1$, with limit

$$\begin{cases} \frac{1}{k} & \text{if } j = k - 1 \\ 0 & \text{if } j < k - 1 \end{cases} \tag{3.39}$$

In particular, it diverges for $k = 0$. For $\lambda = e^{i\theta} \neq 1$, we may use Leibnitz's Rule to obtain

$$[C_n^{\,k}(\lambda)]^{(j)} = [(\lambda^n - 1)(\lambda - 1)^{-1}/n^k]^{(j)} \tag{3.40}$$

$$= \frac{1}{n^k} \sum_{r=0}^{j} \binom{j}{r} n(n-1) \ldots (n-r+1)\lambda^{n-r}(-1)^{j-r}(j-r)!(\lambda - 1)^{-j+r-1}$$

$$= e^{in\theta} n^{j-k} \sum_{r=0}^{j} \frac{n(n-1)\ldots(n-r+1)}{n^j} e^{-i\theta r}(-1)^{j-r}(j-r)!(e^{i\theta}-1)^{-j+r-1}.$$

The largest term in this sum is of order $O(n^{j-k})$ and all terms thus vanish with n, except possibly the term with r = j = k which equals

$$e^{in\theta}\left[\frac{n(n-1) \ldots (n-k+1)}{n^k}\right] \cdot f_k(\theta), \quad f_k(\theta) \text{ bounded.}$$

Hence this term actually diverges with n. Consequently, $C_n^{\,k}(A)$ converges exactly when $|\lambda(A)| \leq 1$, and $|\lambda(A)| = 1 \Rightarrow j \leq k - 1$, that is $\ell \leq k$.

In case $C_n^{\,o}(A)$ converges we see from (ii) that $\rho(A) < 1$ and so $(I - A)^{-1}$ exists, which is well-known to be the actual limit. (ii) $\Leftrightarrow$ (iii) By spectral theorem

$$A^n/n^k \to 0 \Leftrightarrow \left(\frac{\lambda^n}{n^k}\right)_{\lambda_t}^{(j)} \to 0 \quad j = 0, \ldots, i(\lambda_t) - 1$$

$$\Leftrightarrow \frac{n(n-1) \ldots (n-j+1)}{n^k} \lambda_t^{\,n-j} \to 0$$

$$\Leftrightarrow |\lambda(A)| \leq 1, \ |\lambda(A)| = 1 \Rightarrow i(\lambda(A)) \leq k.$$

We note in passing that (3.15a) showed that (i) implies the convergence of A^n/n^k but could not predict the actual limit. Since a necessary condition for the convergence of $C_n^{\,k}(A)$ is that $k \geq \ell$, parts (α) and (β) of Theorem 3.2 are <u>directly applicable</u>. The actual limit of $C_n^{\,k}(A)$ could also be obtained from the spectral representation $C_n^{\,k}(A) = \sum_t \sum_j \sum_{r=0}^{n-1} (\lambda^r)_{\lambda_t}^{(j)} Z_t^{\,j}/n^k$ with aid of (3.38) and (3.39). Indeed, the only non-vanishing term occurs when $\lambda_t = 1$ and $j = \ell - 1 = k - 1$. This yields a limit of $(I - A)^{-1}$, if $k = \ell = 0$, and $\frac{1}{k} Z_1^{\,k-1}$ if $k \geq 1$. This is more compact but less convenient to work with than (ix).

The equivalence of (iii), (iv) and (v) was shown in (3.22). (ii) $\Leftrightarrow$ (vi) This is an immediate consequence of the shift lemma. (ii) $\Leftrightarrow$ (vii) By the shift lemma with $E_1 = A(A - I)(A - I)^d$,

$$\lambda(E_1) = \begin{cases} \lambda(A) & \text{if } \lambda(A) \neq 1 \\ 0 & \text{if } \lambda(A) = 1, \end{cases}$$

and hence (ii) is equivalent to $k \geq \ell$ and

$$|\lambda(E_1)| \leq 1, \quad |\lambda(E_1)| = 1 \Rightarrow i(\lambda(E_1)) \leq k,$$

which by (i) and (ii) is equivalent to (vii). (vii) $\Leftrightarrow$ (viii) By the equivalence of (i) and (iii), $C_n^{\ k}(E_1)$ converges $\Leftrightarrow$ $A^n(A - I)(A - I)^d/n^k$ converges to zero $\Leftrightarrow$ $A^n(A - I)^d/n^k$ converges to zero. □

Two immediate consequences are the following in which the above notation is used.

Corollary 3.9

(i) $C_n^{\ k}(A) \to 0 \Leftrightarrow k \geq 1,\ i(A - I) \leq k = 1$ and $\qquad$ (3.41)

$|\lambda(A)| \leq 1,\ |\lambda(A)| = 1 \Rightarrow i(\lambda(A)) \leq k$

(ii) $C_n^{\ k}(A_1) \to 0$ for some $(A - I)^- \Leftrightarrow k \geq 1$ and

$|\lambda(A)| \leq 1,\ |\lambda(A)| = 1 \Rightarrow i(\lambda(A) \leq k$

$\Leftrightarrow C_n^{\ k}(A_1) \to 0$ for all $(A - I)^-$.

Proof (i) This follows immediately from (3.37)ix and the fact that $Z_1^{\ k-1}(A) = 0 \Leftrightarrow \ell \leq k - 1$. Alternatively, (3.39) could be used for this. (ii) Use (3.15a) and (3.37)ii. □

Remarks 1. Part (i) shows with aid of (3.37) that

$$C_n^{\ k}(A) \to 0 \Leftrightarrow k \geq 1 \text{ and } A^n/n^k \to 0,\ i(A - I) < k - 1 \tag{3.42}$$

2. If for some $k \geq 1$, $C_n^{\ k-1}(A)$ diverges and $C_n^{\ k}(A)$ converges, then there is an eigenvalue λ_o such that $|\lambda_o| = 1$, and $i(\lambda_o) = k$. If in addition $C_n^{\ k}(A) \to 0$ then $i(1) \leq k - 1$ and hence $\lambda_o \neq 1$.

Theorem 3.10 Let $\ell = i(A - I)$. The following are equivalent.

(i) A^n/n^k converges. (ii) $C_n^{\ k}(A_1)$ converges. (iii) $A_1^{\ n}/n^k \to 0$.

(iv) $|\lambda(A_1)| \leq 1,\ |\lambda(A_1)| = 1 \Rightarrow i(\lambda(A_1)) \leq k$.

(v) $\left.\begin{array}{l}|\lambda(A)| \leq 1, \; |\lambda(A)| = 1 \\ \qquad\qquad\quad \lambda(A) \neq 1\end{array}\right\} \Rightarrow i(\lambda(A)) \leq k, \; i(A - I) \leq k + 1.$

(vi) $|\lambda(E_1)| \leq 1, \; |\lambda(E_1)| = 1 \Rightarrow i(\lambda(E_1)) \leq k, \; i(A - I) \leq k + 1.$

(vii) $C_n^{\,k}(E_1)$ converges and $i(A - I) \leq k + 1$.

(viii) $A^n(A - I)^d/n^k \to 0$ and $i(A - I) \leq k + 1$.

In which case $A^n/n^k \to Z_1^{\,k}(A)$ and

$$\text{(ix)} \quad C_n^{\,k}(A_1) \to \begin{cases} (I - A)^{\#}(A - I)(A - I)^- + I - (A - I)(A - I)^- & \text{if } k = 0 \\ \frac{1}{k}\,[Z_1^{\,k-1}(A)](A - I)(A - I)^- & \text{if } k \geq 1 \end{cases}$$

Proof (i) $\Leftrightarrow$ (ii) This follows from (3.15). (ii) $\Leftrightarrow$ (iii) $\Leftrightarrow$ (iv) An immediate consequence of Theorem 3.8. The equivalence of (iv), (v) and (vi) follows again from the shift lemma, while the equivalence of (vi), (vii) and (viii) follows from (3.37) applied to E_1. Next,

$$\lim_n A^n/n^k = \lim_n \sum_{t=1}^{s} \sum_{j=0}^{m_t - 1} \frac{n(n - 1) \ldots (n - j + 1)}{n^k} \lambda_t^{\,n-j} Z_t^{\,j},$$

in which the only non-zero terms come from $\lambda_t = 1$, if any, and $j = k = \ell - 1$, and yield $Z_1^{\,\ell-1}(A)$. For $k = 0$, the convergence of A^n implies that $\ell \leq 1$ and so this reduces to the well-known result: $A^n \to Z_1^{\,o} = I - (A - I)(A - I)^{\#}$. Lastly, by Theorem (3.8)ix,

$$\lim C_n^{\,k}(A_1) = \begin{cases} (I - A_1)^{-1} & \text{if } k = i(A_1 - I) = 0 \\ \frac{1}{k} Z_1^{\,k-1}(A_1) & \text{if } k \geq 1, \; k \geq i(A_1 - I) \end{cases}$$

We may now use Theorem (3.1)vii and Theorem (4.6)xv to express these limits in terms of A. Indeed for $k = 0 = i(A_1 - I)$, $i(A - I) = 1$ and $(I - A_1)^{-1} = (X^2X^- + I - XX^-)^{-1} = X^{\#}XX^- + I - XX^-$, where $X = I - A$. Similarly for $k \geq 1$ and all $j = 0,1,2,\ldots$

$$\begin{aligned} Z_1^{\,j}(A_1) &= (A_1 - I)^j[I - (A_1 - I)(A_1 - I)^d]/j! \\ &= (-1)^j(X^{j+1}X^- + I - XX^-)[I - (X^2X^- + I - XX^-)(X^dXX^- + I - XX^-)]/j! \\ &= (-1)^jX^j(I - XX^d)XX^-/j! = Z_1^{\,j}(A) \cdot (A - I)(A - I)^-, \end{aligned}$$

which yields the final result.

We may now combine the above results (3.37)-(3.43) to conclude that:

(i) for $k \geq \ell$, A^n/n^k converges $\Leftrightarrow$ $A^n/n^k \to 0$ and $C_n^{\ k}(A)$ converges $\Leftrightarrow$ $C_n^{\ k}(A_1)$ converges. Moreover the limits (3.22)βiii and (3.42)a agree.

(ii) for $k \geq \ell - 1$, $A_1^{\ n}/n^k$ converges $\Leftrightarrow$ $A_1^{\ n}/n^k \to 0 \Leftrightarrow A^n/n^k$ converges.

(iii) $C_n^{\ k}(A) \to 0 \Leftrightarrow C_n^{\ k}(A)$ converges and $k \geq \ell + 1$. □ (3.44)

In table form we may recapitulate these as

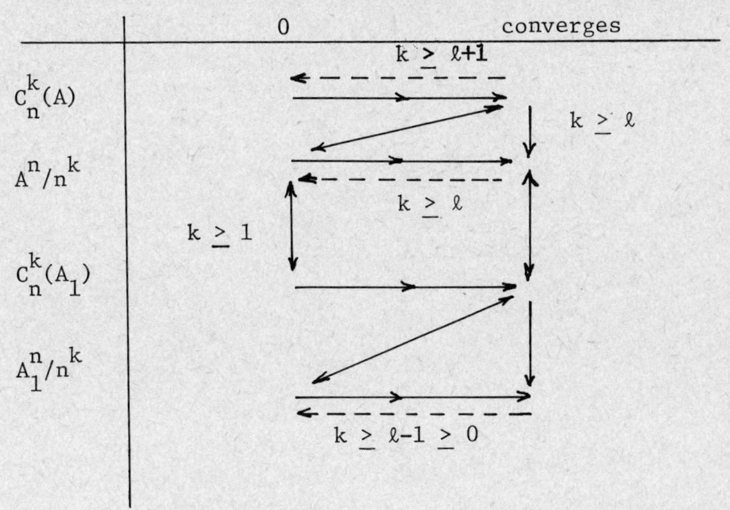

Let us now turn to the last part of this section and consider the convergence behavior of $D_n^{\ k}(A)$. We may in fact use most of the last few results, once we recall the useful identities (3.17). Unlike with $C_n^{\ k}(A)$, Theorem 3.2 did not give the complete picture.

Theorem 3.11 Let $i(A - I) = \ell$. The following are equivalent for $k \geq 0$.

(i) $D_n^{\ k}(A_1)$ converges. (ii) $AC_n^{\ k}(A)/n$ converges. (3.45)

(iii) $C_n^{\ k+1}(A)$ converges. (iv) $A^n/n^{k+1} \to 0$

(v) $|\lambda(A)| \leq 1$, $|\lambda(A)| = 1 \Rightarrow i(\lambda(A)) \leq k + 1$.

(vi) $|\lambda(A_1)| \leq 1$, $i(A_1 - I) \leq k$, $|\lambda(A_1)| = 1 \Rightarrow i(\lambda(A_1)) \leq k + 1$.

In which case

$$D_n^{\,k}(A_1) \to \begin{cases} (I - A)^{-1} & \text{if } k = \ell = 0 \\ 0 & \text{if } k > \ell = 0 \\ Z_1^{\,o}(A) \cdot (A - I)^- + I - A(A - I)^- & \text{if } k = 0,\ \ell = 1 \\ \frac{1}{\ell} Z_1^{\,\ell-1}(A) \cdot (A - I)^- & \text{if } k + 1 = \ell \geq 1,\ k \geq 1 \\ 0 & \text{if } k + 1 > \ell \geq 1. \end{cases} \tag{3.46}$$

<u>Proof</u> The equivalence of (i)-(iii) follows at once from Theorem 3.2 and the fact that $C_n^{\,k+1}(A)$ converges (with the same limit!). (iii) $\Leftrightarrow$ (iv) $\Leftrightarrow$ (v) Clear by (3.37). (v) $\Leftrightarrow$ (vi) This is a direct consequence of the shift lemma.

To compute $\lim_n D_n^{\,k}(A_1)$ we first note that we cannot use (3.22)δii, since all we know is that $k \geq \ell - 1$. However from (3.16)b and (3.17)a we arrive at

$$\begin{aligned} \lim_n D_n^{\,k}(A_1) &= [\lim_n \tfrac{A}{n} C_n^{\,k}(A)](A - I)^- + [I - A(A - I)^-]\delta_{k,o}/n^k \\ &= [\lim_n C_n^{\,k+1}(A)](A - I)^- + [I - A(A - I)^-]\delta_{k,o}/n^k. \end{aligned} \tag{3.47}$$

Now from (3.37)ix we see that the case $k + 1 = \ell = 0$ cannot occur and furthermore easily obtain (vii). □

<u>Corollary 3.12</u> Let $i(A - I) = \ell$. The following are equivalent for $k \geq 0$.

(i) $D_n^{\,k}(A)$ converges. (3.48)

(ii) $|\lambda(A)| \leq 1$, $|\lambda(A)| = 1 \Rightarrow i(\lambda(A)) \leq k + 1$, $i(A - I) \leq k$.

(iii) $|\lambda(A_1)| \leq 1$, $|\lambda(A_1)| = 1 \Rightarrow i(\lambda(A_1)) \leq k + 1$, $i(A_1 - I) \leq k - 1$.

In which case

$$D_n^{\,k}(A) \to \begin{cases} (I - A)^{-1} & \text{if } k = \ell = 0 \\ \frac{1}{(k + 1)k} Z_1^{\,k-1}(A) & \text{if } k \geq 1. \end{cases}$$

<u>Proof</u> We begin by noting that one may not in general replace A_1 by A in (3.47)(i) and (vi) because <u>not</u> all matrices A are of the form

$B(B - I)(B - I)^-$ for some B. (ii) ⇒ (i) This follows at once from Theorem (3.2). (i) ⇒ (ii) If $D_n^k(A)$ converges so does $D_n^k(A_1)$ and hence by (3.47) $A^n/n^{k+1} \to 0$ and $|\lambda(A)| \leq 1$, $|\lambda(A)| = 1 \Rightarrow i(\lambda(A)) < k + 1$. In particular $\ell = i(A - I) \leq k + 1$. Now on examining (3.28) we see that all but the last term also converge, forcing $\ell \leq k$. (ii) ⇔ (iii) Use the shift lemma. To compute $\lim D_n^k(A)$, all we need to recall is that a necessary condition for its convergence was that $k \geq \ell$, so that (3.22)δv is immediately applicable. □

Remarks 1. For $k \geq \ell$, $D_n^k(A)$ converges exactly when $D_n^k(A_1)$ converges and the limiting expressions (3.45)vii and (3.22)δii agree.

2. From the limiting expressions for C_n^k and D_n^k we see that if they converge then:

$$\lim_n D_n^k(A) = \frac{1}{k+1} \lim_n C_n^k(A), \quad k \geq 0. \tag{3.49}$$

3. In the above convergence theorems we may replace A_1 by the similar matrix $B_1 = (A - I)^-(A - I)A$.

4. If $\ell = 1$, then $\lim C_n^o(A_1) = \lim D_n^o(A_1) = (I - A)^{\#}(A - I)(A - I)^- + I - (A - I)(A - I)^-$, provided $|\lambda(A)| < 1$ if $\lambda(A) \neq 1$.

5. Clearly, $D_n^k(A) \to 0 \Leftrightarrow D_n^k(A)$ converges and $k \geq \ell + 1$.

6. If one uses the spectral theorem to establish the convergence behavior of $D_n^k(A)$, one has to deal with expressions similar to those given in (3.28).

4. THE GOVERNING MATRICES

In this section we shall derive master theorems for the matrices U_1, V_1, P and Q whose spectra determine in the complex case the convergence behavior of S_n^k, T_n^k. The case of zero index is somewhat special and will be treated separately, because of the fact that $i(M) = 0 \Leftrightarrow i(M^r) = 0$, $r \geq 1$, which is not true for larger indices in general. Our main theorems will be slightly more general than needed and will deal with the matrices

$$\begin{aligned} U &= XAXX^{=} + I - XX^- & V &= X^{=}XAX + I - X^-X \\ P &= XA + I - XX^- & Q &= AX + I - X^-X, \end{aligned} \tag{4.1}$$

where $A \in \mathcal{R}_{m\times m}$, $X \in \mathcal{R}_{n\times m}$, and $X^-, X^=$ are two possibly distinct inner inverses of X. U_1 and V_1 are obtained from these on setting $X^= = X^-$.

As always, $\mathcal{R}$ will denote a regular ring with unity, and we use the usual convention that $\sum_1^{k-1}$ is empty when k = 1. We begin with several basic results.

Theorem 4.1 Let U, U_1, P, V, V_1 and Q be given as in (4.1). Then for all $k \geq 0$,

(i) $$U^k = (XA)^k XX^- + [(XA)^k + \ldots + XA]XX^=(I - XX^-) + I - XX^- \quad (4.2)$$
$$= (XA)^k XX^= + [(XA)^{k-1} + \ldots + XA]XX^=(I - XX^-)$$
$$+ I - XX^- \quad (k \geq 1)$$

(ii) $$U_1^k = (XA)^k XX^- + I - XX^-$$

(iii) $$P^k = (XA)^k XX^- + [(XA)^k + \ldots + XA + I](I - XX^-)$$
$$= (XA)^k + [(XA)^{k-1} + \ldots + XA + I](I - XX^-) \quad (k \geq 1)$$

(iv) $$N(U^k) = N(U_1^k) = N(P^k) = N[(XA)^k] \cap R(X) = N[(XA)^k XX^=] \cap R(X)$$
$$= N[(XA)^k XX^= + ((XA)^{k-1} + \ldots + XA)XX^=(I - XX^-)] \cap R(X)$$
$$= N[(XA)^k + ((XA)^{k-1} + \ldots + XA)(I - XX^-)] \cap R(X)$$

(v) $$RS(U^k) = RS(U_1^k) = RS(P^k) = RS(XA)^k + RS(I - XX^-)$$
$$= RS[(XA)^k XX^-] \oplus RS(I - XX^-) = RS[(XA)^k XX^=$$
$$+ \{(XA)^{k-1} + \ldots + XA\}XX^=(I - XX^-)] + RS(I - XX^-)$$
$$= RS[(XA)^k XX^=] + RS[\{(XA)^{k-1} + \ldots + XA\}XX^=(I - XX^-)$$
$$+ (I - XX^-)] \text{ and } \ell(U) = \ell(U_1) = \ell(P)$$

(vi) $$^o(U^k) = {}^o[(XA)^k X] \cap {}^o[\{(XA)^{k-1} + \ldots + XA\}XX^=(I - XX^-)$$
$$+ I - XX^-]$$

(vii) $$^o(U_1^k) = {}^o[(XA)^k X] \cap {}^o(I - XX^-)$$

(viii) $$^o(P^k) = {}^o[(XA)^k X] \cap {}^o[\{(XA)^k + \ldots + XA + I\}(I - XX^-)]$$
$$\supseteq {}^o[(XA)^k] \cap {}^o[\{(XA)^{k-1} + \ldots + XA + I\}(I - XX^-)]$$
$$\supseteq {}^o(P^{k-1})$$

(ix) $R(U^k) = R[(XA)^k X] + R[\{(XA)^{k-1} + \ldots + XA\}XX^{=}(I - XX^{-})$

$+ I - XX^{-}]$

(x) $R(U_1{}^k) = R[(XA)^k X] + R(I - XX^{-})$

(xi) $R(P^k) = R[(XA)^k X] + R[\{(XA)^k + \ldots + XA + I\}(I - XX^{-})]$

$\subseteq R[(XA)^k] + R[\{(XA)^{k-1} + \ldots + XA + I\}(I - XX^{-})]$

$\subseteq R(P^{k-1})$

(xii) $U_1{}^k P = (XA)^{k+1} + I - XX^{-}$

(xiii) $\ell(I - U_1) \leq \ell(I - P) \leq \ell(I - U_1) + 1$

Analogous results (i)'-(xiii)' involving rowspaces and left annihilators hold for the matrices V, V_1 and Q.

<u>Proof</u> The identities (i)-(iii) follow by induction. Note that $P^k = U_1{}^k + [(XA)^k + \ldots + XA](I - XX^{-})$. (iv) $U^k \underset{\sim}{y} = \underset{\sim}{0} \Rightarrow XX^{-}U^k \underset{\sim}{y} = 0 \Rightarrow [(XA)^k XX^{=} + \{(XA)^{k-1} + \ldots + XA\}XX^{=}(I - XX^{-})]\underset{\sim}{y} = \underset{\sim}{0}$. Combining these, $(I - XX^{-})\underset{\sim}{y} = \underset{\sim}{0} \Rightarrow XX^{=}\underset{\sim}{y} = \underset{\sim}{y} \Rightarrow (XA)^k XX^{=}\underset{\sim}{y} = (XA)^k \underset{\sim}{y} = \underset{\sim}{0}$. Hence $\underset{\sim}{y} \in N[(XA)^k] \cap R(X)$ and $\underset{\sim}{y} \in N[(XA)^k XX^{=}] \cap R(X)$. Conversely, if $(XA)^k \underset{\sim}{y} = \underset{\sim}{0}$ and $\underset{\sim}{y} = XX^{-}\underset{\sim}{y} = XX^{=}\underset{\sim}{y}$ then $U^k \underset{\sim}{y} = \underset{\sim}{0}$. Similarly, $(XA)^k XX^{=}\underset{\sim}{y} = \underset{\sim}{0}$ and $XX^{=}\underset{\sim}{y} = \underset{\sim}{y}$ also imply that $U^k \underset{\sim}{y} = \underset{\sim}{0}$. The third representation follows easily from the above. Next, $P^k \underset{\sim}{y} = \underset{\sim}{0} \Rightarrow XX^{-}P^k \underset{\sim}{y} = \underset{\sim}{0} \Rightarrow [(XA)^k + \{(XA)^{k-1} + \ldots + I\}(I - XX^{-})]\underset{\sim}{y} = \underset{\sim}{0} \Rightarrow (I - XX^{-})\underset{\sim}{y} = \underset{\sim}{0} \Rightarrow \underset{\sim}{y} \in R(X) \Rightarrow (XA)^k \underset{\sim}{y} = \underset{\sim}{0}$.

The converse is obvious. The last representation is now also clear. Since $N(U^k)$ is <u>independent</u> of the choice of $X^{=}$ we may take $X^{=} = X^{-}$ yielding $N(U^k) = N(U_1{}^k)$. (v) This is the dual of (iv). Over a regular ring it is always true that [12] $N[{}^oA \cap {}^oB] = R(A) + R(B)$, ${}^o[N(A) \cap N(B)] = RS(A) + RS(B)$ where $N(\cdot)$ and ${}^o(\cdot)$, denote left and right annihilators of $(\cdot)$, so that $RS(A) = RS(B) \Leftrightarrow N(A) = N(B)$. (vi) $\underset{\sim}{y}^T U^k = \underset{\sim}{0}^T \Rightarrow \underset{\sim}{y}^T U^k XX^{-} = \underset{\sim}{0}^T \Rightarrow \underset{\sim}{y}^T (XA)^k XX^{=} = \underset{\sim}{0}^T \Rightarrow \underset{\sim}{y}^T (XA)^k X = \underset{\sim}{0}^T \Rightarrow \underset{\sim}{y}^T\{[(XA)^{k-1} + \ldots + XA]XX^{=}(I - XX^{-}) + I - XX^{-}\} = \underset{\sim}{0}^T \Rightarrow \underset{\sim}{y}^T$ is contained in the right hand side. The converse is easy. (vii) Set $X^{=} = X$ in (vi). (viii) If $\underset{\sim}{y}^T P^k = \underset{\sim}{0}^T$, then $\underset{\sim}{0}^T = \underset{\sim}{y}^T P^k XX^{-} = \underset{\sim}{y}^T (XA)^k XX^{-} \Rightarrow \underset{\sim}{y}^T[(XA)^k + \ldots + I](I - XX^{-}) = \underset{\sim}{0}^T \Rightarrow \underset{\sim}{y}^T \in {}^o[(XA)^k X] \cap {}^o[\{(XA)^k + \ldots + I\}(I - XX^{-})]$. The converse is clear. Now by the above, if $\underset{\sim}{y}^T P^{k-1} = \underset{\sim}{0}^T$, then $\underset{\sim}{y}^T (XA)^{k-1} X = \underset{\sim}{0}^T = \underset{\sim}{y}^T[(XA)^{k-1} + \ldots + I](I - XX^{-})$. Hence also $\underset{\sim}{y}^T (XA)^k = \underset{\sim}{0}^T = \underset{\sim}{y}^T (XA)^k X$ and $\underset{\sim}{y}^T[(XA)^k + \ldots XA + I](I - XX^{-}) = \underset{\sim}{0}$.

(ix)-(xi) These are the duals of (vi)-(vii). (xii) Obvious when multiplied out. (xiii) From (2.9)(i) we see that

$$r(I - U_1) = k \Leftrightarrow R[(I - XA)^{k+1}X] = R[(I - XA)^{k}X]$$

$$\ell(I - U_1) = k \Leftrightarrow RS[(I - XA)^{k+1}X] = RS[(I - XA)^{k}X].$$

Hence from (2.9)(iii) if $\ell(I - U_1) = k$, then $RS(I - P)^{k+1}$ $= RS[(I - XA)^{k}XX^{-}(I - XA)] = RS[(I - XA)^{k+1}XX^{-}(I - XA)] = RS(I - P)^{k+2}$, and so $\ell(I - P) \leq k + 1$. Conversely, if $\ell(I - P) = \ell$, then $RS[(I - XA)^{\ell+1}X]$ $= RS[(I - XA)^{\ell}XX^{-}(I - XA)X] = RS[(I - XA)^{\ell-1}XX^{-}(I - XA)X] = RS[(I - XA)^{\ell}X]$, or $\ell(I - U_1) \leq \ell$. □

We now state our main "one-sided" theorem.

<u>Theorem 4.2</u> Let U, U_1, V, V_1, P and Q be defined as in (4.1). The following are equivalent for all $k \geq 0$.

$$(\alpha)(i) \quad R[(XA)^{k+1}X] = R[(XA)^{k}X] \tag{4.3}$$

(ii) $R[(XA)^{k+2}] = R[(XA)^{k+1}] = R[(XA)^{k}X]$ and

$R[(AX)^{k+2}] = R[(AX)^{k+1}]$

(iii) $R[(XA)^{k+2}] = R[(XA)^{k+1}] = R[(XA)^{k}X]$

(iv) $r(U) \leq k$

(v) $r(U_1) \leq k$

(vi) $r(V) \leq k$

(vii) $r(V_1) \leq k$

(viii) $r(Q) \leq k$

In which case, in addition to (4.2),

(ix) $R(P^{k}) = R(P^{k+1}) = R[(XA)^{k+1}] + R[\{(XA)^{k} + \ldots + XA$
$+ I\}(I - XX^{-})]$

Moreover, if $(XA)^{k+1}XY = (XA)^{k}X$, $k \geq 0$ and if $S = XYX^{-} + I - XX^{=}$ $S_1 = XYX^{-} + I - XX^{-}$, $T_{k+1} = X^{-}X(AX)^{k}Y^{k+1} + (I - X^{=}X)X^{-}X(AX)^{k}$ $(Y^{k} + \ldots + Y + I) + (I - X^{-}X)X^{=}X\{(AX)^{k} + \ldots + AX + I\} + I - X^{-}X,$

$K = XYX^- + I - XYA$ and $L = X^=XY + I - AXY$,

where the same choices of $X^-, X^=$ are used as in (4.1), then

(x) $\quad U^{k+1}S = U^k$, $U_1^{k+1}S = U_1^k$, $V^{k+1}T_{k+1} = V^k$, $V_1^{k+1}T_1 = V_1^k$,

$P^{k+1}K = P^k$ and $Q^{k+1}L = Q^k$.

Analogous results β(i)-β(x) involving nullspaces and left indices, hold for the matrices U, U_1, V, V_1, Q and P.

Proof (α)(i) $\Rightarrow$ (ii) $R(AX)^{k+2} = AR[(XA)^{k+1}X] = AR[(XA)^kX] = R(AX)^{k+1}$, while $R[(XA)^kX] = R[(XA)^{k+1}X] \subseteq R[(XA)^{k+1}] \subseteq R[(XA)^kX]$. Next, $R[(XA)^{k+2}]$ $= XAR(XA)^{k+1} = XAR[(XA)^kX] = R[(XA)^{k+1}X] = R[(XA)^kX]$. (ii) $\Rightarrow$ (iii) Clear. (iii) $\Rightarrow$ (i) Observe that $R[(XA)^kX] = R[(XA)^{k+2}] \subseteq R[(XA)^{k+1}X] \subseteq R[(XA)^kX]$.

Suppose we now replace (i) by

$$(XA)^{k+1}XY = (XA)^kX \quad \text{or} \quad X(AX)^{k+1}Y = X(AX)^k, \tag{4.4}$$

which ensures that $(AX)^{k+2}Y = (AX)^{k+1}$. We then have: (i) $\Rightarrow$ (iv): $U^{k+1}S$ $= [(XA)^{k+1}XX^= + \{(XA)^k + \ldots + XA\}XX^=(I - XX^-) + (I - XX^-)](XYX^- + I - XX^=)$ $= (XA)^{k+1}XYX^- + [(XA)^k + \ldots + XA]XX^=(I - XX^-) + I - XX^- = (XA)^kXX^-$ $+ (XA)^k(XX^= - XX^-) + [(XA)^{k-1} + \ldots + XA]XX^=(I - XX^-) + I - XX^- = U^k$.
(iv) $\Rightarrow$ (v) Set $X^= = X^-$. This, incidentally, also follows from (4.2)ii.
(v) $\Rightarrow$ (i) Note that $R[(XA)^{k+1}X] = XX^-R(U_1)^{k+1} = XX^-R(U_1)^k = R[(XA)^kX]$.
(i) $\Rightarrow$ (vi) First observe that (4.4) implies

$$(XA)^{k+n}XY^n = (XA)^kX = X(AX)^k, \quad n = 1,2,\ldots \tag{4.5}$$

Now $V^{k+1}T_{k+1} = X^=X(AX)^{2k+1}Y^{k+1} + (I - X^-X)X^=X[(AX)^{2k} + \ldots + (AX)^{k+1}]Y^{k+1}$ $+ (I - X^=X)X^-X(AX)^k(Y^k + \ldots + Y + I) + (I - X^-X)X^=X[(AX)^k + \ldots + AX + I]$ $+ I - X^-X = X^=X(AX)^k + (I - X^-X)X^=X(AX)^k(Y + Y^2 + \ldots + Y^k)$ $+ (I - X^=X)X^-X(AX)^k(Y^k + \ldots + Y + I) + (I - X^-X)X^=X[(AX)^k + \ldots + AX + I]$ $+ (I - X^-X) = X^=X(AX)^k + (I - X^=X)X^-X(AX)^k + (I - X^-X)X^=X[(AX)^k + \ldots + I]$ $+ I - X^-X = V^k$. (vi) $\Rightarrow$ (vii) Set $X^= = X^-$ or use (4.2)xi'.
(vii) $\Rightarrow$ (i) Note that $R[(XA)^{k+1}X] = XR(V_1)^{k+1} = XR(V_1)^k = R[(XA)^kX]$.
(i) $\Rightarrow$ (viii) $Q^{k+1}L = [(AX)^{k+1} + (I - X^-X)\{(AX)^k + \ldots + I\}](X^-XY + I - AXY)$ $= (AX)^k + (I - X^-X)[(AX)^{k-1} + \ldots + AX + I] = Q^k$. (viii) $\Rightarrow$ (i) $R[X(AX)^{k+1}]$ $= XR(Q^{k+1}) + XR(Q^k) = R[X(AX)^k]$. (i) $\Rightarrow$ (ix) To prove the last part, it suffices to verify that $P^{k+1}K = [(XA)^{k+1} + \{(XA)^k + \ldots + XA + I\}$.

$(I - XX^-)](XYX^- + I - XYA) = (XA)^{k+1}XYX^- + [(XA)^k + \ldots + I](I - XX^-)(I - XYA) = (XA)^k + [(XA)^{k-1} + \ldots + XA + I](I - XX^-) = P^k$. A final application of (4.2)xi completes the proof. □

For the case k = 0, we may add several additional conditions to the above.

Corollary 4.4 The following are equivalent

(i) $R(XAX) = R(X)$ (4.6)

(ii) $R(XA)^2 = R(XA) = R(X)$, and $R(AX)^2 = R(AX)$

(iii) $R(XA)^2 = R(XA) = R(X)$

(iv) U has a right inverse

(v) U_1 has a right inverse

(vi) V has a right inverse

(vii) V_1 has a right inverse

(viii) P has a right inverse and $R(XA)^2 = R(XA)$

(ix) Q has a right inverse

(x) $(XA)^{\ell+1} + I - XX^-$ has a right inverse for some, and hence all, ℓ.

(xi) $(AX)^{\ell+1} + I - X^-X$ has a right inverse for some, and hence all, ℓ.

(xii) $(XA)^{\ell}XX^- + I - XX^-$ has a right inverse for some, and hence all, ℓ.

(xiii) $X^-X(AX)^{\ell} + I - X^-X$ has a right inverse for some, and hence all, ℓ.

In which case all ranges in (4.2)ix-xi, equal $\mathcal{R}^n$, and those in (v') equal $\mathcal{R}^m$.

The left hand analogue of this is self-evident and is omitted.

Proof (viii) ⇒ (iii) Observe that $R(P) = \mathcal{R}^n \Rightarrow R(XA) = XX^- R(P) = R(X)$. For (x) and (xi) we use (4.2)xii,xii' and recall that $U_1^{\ell+1}P$ has a right inverse if and only if U_1 and P both have right inverses.

Suppose now that $t = \max(m,n)$ and $\mathcal{R}_{t\times t}$ is finite regular, that is one-sided inverses are two-sided inverses, as is the case with commutative and with unit regular rings, then [15] $R(M) = R(N)$, $RS(M) \subseteq RS(N) \Rightarrow RS(M) = RS(N)$, and $RS(M) = RS(N)$, $R(M) \subseteq R(N) \Rightarrow R(M) = R(N)$. This shows that $\ell(M) = k \Rightarrow r(M) = k$, and hence that (4.3)α and (4.3)β are equivalent, as recapitulated in Corollary 4.7. In addition it would then follow that $i(I - U_1) = \min_k \{ R[(I - XA)^{k+1}X] = R[(I - XA)^k X]\} = i(I - V_1)$. □

Moreover, (4.2)xiii becomes $i(I - U_1) \leq i(I - P) \leq i(I - U_1) + 1$, which is sharper than Cline's universal bounds [3], $|i(MN) - i(NM)| \leq 1$ with $M = I - XA$ and $N = XX^-$. Since $I - U_1 = X(I - V_1)X^-$, $I - V_1 = X^-(I - U_1)X$, it follows that for, say, $m \geq n$, $\begin{bmatrix} X \\ 0 \end{bmatrix}(I - V_1)[X^-,0] = \begin{bmatrix} I - U_1 & 0 \\ 0 & 0 \end{bmatrix}_{m\times m}$ and $[X^-,0]\begin{bmatrix} I - U_1 & 0 \\ 0 & 0 \end{bmatrix}\begin{bmatrix} X \\ 0 \end{bmatrix} = I - V_1$. Thus $I - V_1$ and $\begin{bmatrix} I - U_1 & 0 \\ 0 & 0 \end{bmatrix}$ are pseudo-similar. If now $\begin{bmatrix} X \\ 0 \end{bmatrix}$ is unit-regular, then by Lemma (2.3)ii

$$I - V_1 \approx \begin{bmatrix} I - U_1 & 0 \\ 0 & 0 \end{bmatrix} \quad \text{or} \quad V_1 \approx \begin{bmatrix} U_1 & 0 \\ 0 & I_{m-n} \end{bmatrix}.$$

For $m \leq n$ we obviously interchange U_1 with V_1. In particular if $m = n$ and if X is unit regular, then $XAXX^- + I - XX^- = U_1 \approx V_1 = X^-XAX + I - X^-X$. For matrices over a field $\mathcal{F}$, we may combine for $m \geq n$ the above remarks with the lesser known facts [21, p. 164]

$$\Delta_{AB}(\lambda) = \lambda^{m-n}\Delta_{BA}(\lambda),\ \psi_{AB}(\lambda) = \lambda^{o,\pm 1}\psi_{BA}(\lambda),\ A,B^T \in \mathcal{F}_{m\times n},$$

$$\Delta_{I-A}(\lambda) = \Delta_{I-B}(\lambda) \Leftrightarrow \Delta_A(\lambda) = \Delta_B(\lambda),$$

$$\psi_{I-A}(\lambda) = \psi_{I-B}(\lambda) \Leftrightarrow \psi_A(\lambda) = \psi_B(\lambda),$$

$$\psi_{I-A}(\lambda) = \lambda\psi_{I-B}(\lambda) \Leftrightarrow \psi_A(\lambda) = (\lambda - 1)\psi_B(\lambda);$$

to conclude that

(i) $\Delta_{I-V_1}(\lambda) = \lambda^{m-n}\Delta_{I-U_1}(\lambda) = \Delta_{I-Q}(\lambda) = \lambda^{m-n}\Delta_{I-P}(\lambda)$ (4.7)

(ii) $\psi_{I-U_1}(\lambda) = \psi_{I-V_1}(\lambda)$

(iii) $\psi_{I-P}(\lambda) = \lambda^{o,1}\psi_{I-U_1}(\lambda)$, $\psi_{I-Q}(\lambda) = \lambda^{o,1}\psi_{I-V_1}(\lambda)$

(iv) $\Delta_{V_1}(\lambda) = (\lambda - 1)^{m-n}\Delta_{U_1}(\lambda) = \Delta_Q(\lambda) = (\lambda - 1)^{m-n}\Delta_P(\lambda)$

(v) $\psi_P(\lambda) = (\lambda - 1)^{o,1}\psi_{U_1}(\lambda)$, $\psi_Q(\lambda) = (\lambda - 1)^{o,1}\psi_{V_1}(\lambda)$.

These will be used repeatedly in section 5. Before combining (4.3)α and (4.3)β let us first state on the basis of Lemma 2.2 and Corollary 2.6 the following result.

Corollary 4.5

(i) If $R(XA) = R(X)$, then $I - U_1 \approx (I - XA)XA(XA)^-$ (4.8)

(ii) If $RS(AX) = RS(X)$, then $I - V_1 \approx (AX)^-AX(I - AX)$

(iii) If $R(XA) = R(X)$ and $(XA)^\#$ exists, then

$$[I - (XA)^\#]XX^- \approx [I - (XA)^\#]XA(XA)^- \approx [I - (XA)^\#]XA(XA)^\#$$

(iv) If $RS(AX) = RS(X)$ and $(AX)^\#$ exists, then

$$X^-X[I - (AX)^\#] \approx (AX)^-AX[I - (AX)^\#] \approx (AX)^\#AX[I - (AX)^\#].$$

We note with aid of the shift lemma, that when $R = F$ and U_1 is invertible, its spectrum is entirely determined by that of XA. This, in fact, is also the case for larger indices. Combining the results of (4.3)α,β we obtain our most useful two-sided theorem of this section.

Theorem 4.6 The following are equivalent for $k \geq 0$.

(i) $R[(XA)^{k+1}X] = R[(XA)^kX]$ and $RS[(XA)^{k+1}X] = RS[(XA)^kX]$ (4.9)

(ii) $(XA)^{k+1}$ and $(AX)^{k+1}$ have group inverses and $R[(XA)^{k+1}]$ $= R[(XA)^kX]$, $RS[(AX)^{k+1}] = RS[X(AX)^k]$

(iii) $i(XA) \leq k + 1$ and $R[(XA)^{k+1}] = R[(XA)^kX]$

(iv) $(XA)^d$ exists and $(XA)^{k+1}(XA)^dX = (XA)^kX$

(v) $i(AX) \leq k + 1$ and $RS[(AX)^{k+1}] = RS[X(AX)^k]$

(vi) $(AX)^d$ exists and $(XA)^k X = X(AX)^d (AX)^{k+1}$

(vii) $i(U) \leq k$

(viii) $i(U_1) \leq k$

(ix) $i(V) \leq k$

(x) $i(V_1) \leq k$

In which case

(xi) $i(P) \leq k$

(xii) $i(Q) \leq k$

Moreover if

$$(*) \quad (XA)^{k+1} XY = (XA)^k X = ZX(AX)^{k+1},$$

then

$$\text{(xiii)} \quad (XA)^k U^d X = (XA)^k XYX^- X = (XA)^k XY = ZX(AX)^k = XX^- ZX(AX)^k$$
$$= XV^d (AX)^k$$

$$\text{(xiv)} \quad Y = (AX)^k [(AX)^{k+1}]^{\sim}, \; Y = (AX)^d, \; Z = [(XA)^{k+1}]^{\sim} (XA)^k,$$
$$Z = (XA)^d$$

are particular solutions, to (*).

$$\text{(xv)} \quad U^d = C_{k+1} = (XA)^k XY^{k+1} X^- + (XA)^k X[Y^k + \ldots + Y + I]X^- (I - XX^=)$$
$$+ [(XA)^k + \ldots + XA + I]XX^= (I - XX^-) + I - XX^=$$
$$= (XA)^d XX^- + [(XA)^k + \ldots + XA + I](I - (XA)^d)XX^= (I - XX^-)$$
$$+ (I - (XA)^d XX^-)(I - XX^=) - (XA)^k XX^= (I - XX^-),$$

$$U_1{}^d = Z^{k+1} (XA)^k XX^- + I - XX^- = X(AX)^k Y^k X^- + I - XX^-$$
$$= (XA)^d XX^- + I - XX^-,$$

$$V^d = T_{k+1} = X^- Z^{k+1} X(AX)^k + (I - X^= X)X^- (Z^k + \ldots + Z + I)X(AX)^k$$
$$+ (I - X^- X)X^= X[(AX)^k + \ldots + AX + I] + I - X^= X = X^- X(AX)^d$$
$$+ (I - X^= X)X^- X(I - (AX)^d)[(AX)^k + \ldots + AX + I]$$
$$+ (I - X^= X)[I - X^- X(AX)^d] - (I - X^- X)X^= X(AX)^k,$$

$$V_1^d = X^- Z^{k+1}(XA)^k X + I - X^- X = X^-(XA)^k XY^{k+1} + I - X^- X$$
$$= X^- X(AX)^d + I - X^- X,$$
$$P^d = (Z^{k+1} + \ldots + Z)(XA)^k XX^-(I - XA) + (XA)^k$$
$$+ [(XA)^{k-1} + \ldots + XA + I](I - XX^-)$$
$$= (XA)^k X(Y^{k+1} + \ldots + Y)X^-(I - XA) + (XA)^k$$
$$+ [(XA)^{k-1} + \ldots + I](I - XX^-)$$
$$= [I + XA + \ldots + (XA)^k][(XA)^d XX^- + I - XA(XA)^d]$$
$$- [I + XA + \ldots + (XA)^{k-1}]XX^-,$$
$$Q^d = (I - AX)X^- X(AX)^k(Y^{k+1} + \ldots + Y) + (AX)^k$$
$$+ (I - X^- X)[(AX)^{k-1} + \ldots + AX + I]$$
$$= (I - AX)X^-(Z^{k+1} + \ldots + Z)X(AX)^k + (AX)^k$$
$$+ (I - X^- X)[(AX)^{k-1} + \ldots + AX + I] = [X^- X(AX)^d$$
$$+ I - AX(AX)^d][I + AX + \ldots + (AX)^k]$$
$$- X^- X[I + AX + \ldots + (AX)^{k-1}].$$

Proof It is clear from (4.3)α,β that (i), (ii) and (iii) are equivalent. (iii) $\Rightarrow$ (iv) Observe that $XA[(XA)^{k+1}(XA)^d X - (XA)^k X] = 0 \Rightarrow R[(XA)^{k+1}(XA)^d X - (XA)^k X] \subseteq N(XA)^{k+1} \cap R[(XA)^k X] = N(XA)^{k+1} \cap R(XA)^{k+1} = \{0\}$. (iv) $\Rightarrow$ (i) Clear, since XA and $(XA)^d$ commute. The equivalence of (i), (v) and (vi) follows by symmetry. Parts (vii) through (xii) follow at once from Theorem (4.3)α,β. (xiii) First observe that $(XA)^k XY = Z(XA)^{k+1}XY = Z(XA)^k X$. Pre-multiplication by XX^- and post-multiplication by $X^- X$ yields the second and fourth equalities. These identities reflect the fact that if $M^{k+1}F = M^k = GM^{k+1}$, then $MM^d = M^k F^k = G^k M^k$. For $k = 0$, this simply states that left and right inverses must be equal. With aid of (xv), (4.5) and (4.6) we have $(XA)^k U^d X = (XA)^k[(XA)^k XY^{k+1}]X^- X = (XA)^k Z^{k+1}(XA)^k XX^- X = (XA)^k(XA)^k(XA)^k XY^{k+1} = (XA)^{2k}XY^k Y = (XA)^k XY$. Similarly for the remaining identity. (xiv) Observe that $A[(XA)^{k+1}XY - (XA)^k X] = (AX)^{k+2}Y - (AX)^{k+1}$. Since $i(AX) \le k + 1$, we may for example take $Y = (AX)^d$ or $(AX)^k[(AX)^{k+1}]^\sim$, because then $R[(XA)^{k+1}XY - (XA)^k X] \subseteq R[(XA)^k X] \cap N(XA)^{k+1} = \{0\}$. Similarly for the choices of Z. (xv) We may either verify directly that $C_{k+1}U = UC_{k+1}$ and $C_{k+1}UC_{k+1} = C_{k+1}$ or use (2.8) which with aid of (xiii) gives:

$U^d = U^kS^{k+1} = [(XA)^kXX^= + \{(XA)^{k-1} + \ldots + XA\}XX^=(I - XX^-) + I - XX^-](XYX^- + I - XX^-)^{k+1} = [Z(XA)^kXX^- + [(XA)^{k-1} + \ldots + XA]XX^=(I - XX^-) + I - XX^-]S^k$, which by induction reduces to $U^d = [Z^\ell X(AX)^kX^- + (Z^{\ell-1} + \ldots + Z)(XA)^kXX^-(I - XX^=) + [(XA)^{k-1} + \ldots + XA]XX^=(I - X) + I - XX^-]S^{k-\ell+1}$, $\ell = 1,\ldots,k + 1$, which again equals C_{k+1} when $\ell = k + 1$. On applying the identity

$$Z^r(XA)^k = (XA)^kXY^r \qquad r = 1,2,\ldots,k \geq 0, \tag{4.10}$$

we obtain the second expression for U^d, while setting $Z = (XA)^d$ easily yields the last representation for U^d. The corresponding results for V^d follow in step by step analogous fashion.

Lastly, $P^d = P^kD^{k+1} = [(XA)^k + \{(XA)^{k-1} + \ldots + XA + I\}(I - XX^-)](XYX^- + I - XYA)^{k+1}$. Now observe that with aid of (xiii),

$$\begin{aligned}[(XA)^kXX^- - (XA)^{k+1}](XYX^- + I - XYA) &= Z[(XA)^kXX^- - (XA)^{k+1}] \\ &= P^kK - P^k\end{aligned} \tag{4.11}$$

and that $(I - XX^-)(I - XYA) = I - XX^-$. Hence by induction:
$P^d = [(Z^\ell + \ldots + Z)(XA)^kXX^-(I - XA) + P^k]K^{k-\ell+1} = [(XA)^kX(Y^\ell + \ldots + Y)X^-(I - XA) + P^k]K^{k-\ell+1}$, $\ell = 1,\ldots,k + 1$. On setting $Z = (XA)^d$ this gives $P^d = (XA)^d[I + XA + \ldots + (XA)^k]XX^-(I - XA) + P^k$ which reduces with aid of (4.1)iii to the last representation of P^d. We remark that with this choice of $Z = (XA)^d$, on application of Cline's formula [3],

$$(AB)^d = A[(BA)^d]^2B \tag{4.12}$$

yields $K = (XA)^dXX + I - XA(XA)^d = E$, which <u>commutes</u> with P. Moreover, by induction

$$K^k = [(XA)^d]^kXX^- - [[(XA)^d]^k + \ldots + (XA)^d](I - XX^-) + I - XA(XA)^d,$$

which may also be used to compute P^d. The representations for Q^d follow by left-right symmetry. □

In the case where $\mathcal{R}_{t\times t}$ is a finite regular ring, as for example with $¢_{t\times t}$, we may simplify the above considerably to the following.

Corollary 4.7 The following are equivalent for $k \geq 0$.

(i) $R[(XA)^{k+1}X] = R[(XA)^k X]$

(ii) $i(XA) \leq k + 1$ and $R[(XA)^{k+1}] = R[(XA)^k X]$

(iii) $i(AX) \leq k + 1$ and $RS(AX)^{k+1} = RS[X(AX)^k]$

(iv) $i(U) \leq k$

(v) $i(U_1) \leq k$

(vi) $i(V) \leq k$

(vii) $i(V_1) \leq k$

(viii) $i(P) \leq k$

(ix) $i(Q) \leq k$

5. THE CESARO-NEUMANN ITERATIONS S_n^k, T_n^k

We shall now combine all the algebraic and analytic results of the previous sections to investigate the convergence behavior and the limit of the complex Cesaro-Neumann iterations:

$$S_n^k = \sum_{r=0}^{n-1} (I - XA)^r X/n^k, \quad \text{and} \tag{5.1}$$

$$T_n^k = \sum_{r=0}^{n-1} (n - r)(I - XA)^r X/n^{k+1}, \quad k \geq 0, \quad n \to \infty.$$

The convergence of these sequences is in fact determined by the spectra of the matrices U_1 and XA. Our key observation is the fact that S_n^k is weakly algebraically related to any of the four Cesaro sums $C_n^k(I - U_1)$, $C_n^k(I - V_1)$, $C_n^k(I - P)$ or $C_n^k(I - Q)$. All of these sums have exactly the same convergence behaviors since the four matrices have by (4.7) identical spectra. The equivalence depends on the identities

$$(I - XA)^r X = (I - U_1)^r X = (I - P)^r X = X(I - V_1)^r = X(I - Q)^r$$
$$= X(I - AX)^r \quad r \geq 0 \quad \text{and}$$

$$(I - U_1)^r = (I - XA)^r XX^-, \quad (I - V_1)^r = X^- X(I - AX)^r, \quad r \geq 1$$

together with (see (2.9)),

$$(I - P)^r = (I - XA)^{r-1}XX^-(I - XA),$$

$$(I - Q)^r = (I - AX)X^-X(I - AX)^{r-1} \qquad r \geq 1,$$

which imply that for all $k \geq 0$,

$$S_n^k = [C_n^k(I - U_1)]X, \quad C_n^k(I - U_1) = S_n^k \cdot X^- + (I - XX^-)/n^k \tag{5.2}$$

$$S_n^k = X \cdot C_n^k(I - V_1), \quad C_n^k(I - V_1) = X^- \cdot S_n^k + (I - X^-X)/n^k$$

$$S_n^k = [C_n^k(I - P)] \cdot X, \quad C_n^k(I - P) = S_n^k \cdot X^-(I - XA)$$
$$+ [I - (I - XA)^{n-1}(I - P)]/n^k$$

$$S_n^k = X \cdot C_n^k(I - Q), \quad C_n^k(I - Q) = (I - AX)X^- \cdot S_n^k$$
$$+ [I - (I - Q)(I - AX)^{n-1}]/n^k.$$

Analogous results hold for T_n^k and $D_n^k(I - (\cdot))$.

Because the iteration involving $V_1(Q)$ is the left analogue of that involving $U_1(P)$, we shall only discuss the latter in detail, leaving the others to symmetry. If one considers S_n^k as being "generated" by, say, the Cesaro expansion $C_n^k(I - U_1)$, then one may equally well consider the "inverse" iteration generated by $C_n^k(I - U_1^d)$. For $k = 0$, these take the particularly simple form

$$U_1^{-1} = [I - (I - XA)XX^-]^{-1} = \sum_{r=0}^{\infty} [(I - XA)XX^-]^r = \lim_n C_n^o(I - U_1) \tag{5.3}$$

$$U_1 = [I - (I - Z)XX^-]^{-1} = \sum_{r=0}^{\infty} [(I - Z)XX^-]^r = \lim_n C_n^o(I - U_1^{-1}). \tag{5.4}$$

which hold for $\max_{\lambda(XA)\neq 0} |1 - \lambda(XA)| < 1$, and $\max_{\lambda(XA)\neq 0} |1 - 1/\lambda(XA)| < 1$ respectively. Similar results hold for the other "inverse" iterations. We shall subsequently use both the iterations in (5.3), (5.4) to show that all types of outer inverses, such as reflexive, Drazin and Moore-Penrose inverses can be generated in this way. First our main theorem, in which we have chosen U_1 to govern our iteration.

Theorem 5.1 The following are equivalent for $k \geq 0$.

(i) S_n^k converges (5.5)

(ii) $S_n^kX^-$ converges

(iii) $C_n^{\,k}(I - U_1)$ converges

(iv) $|1 - \lambda(U_1)| \le 1$, $|1 - \lambda(U_1)| = 1 \Rightarrow i(1 - \lambda(U_1)) \le k$

(v) $(I - U_1)^n/n^k \to 0$

(vi) $(I - XA)^n X/n^k \to 0$

(vii) $[(I - XA)XA(XA)^-]^n/n^k \to 0$ and $i(U_1) \le k$

(viii) $(I - XA)^n/n^k$ converges and $i(U_1) \le k$

(ix) $(I - U_1)^n/n^k$ converges and $i(U_1) \le k$

(x) $C_n^{\,k}[(I - XA)XA(XA)^-]$ converges and $i(U_1) \le k$

(xi) $|\lambda[(I - XA)XA(XA)^-]| \le 1$, $|\lambda[(I - XA)XA(XA)^-] = 1 \Rightarrow i(\lambda) \le k$

and $i(U_1) \le k$

(xii) $\left.\begin{array}{r}|1 - \lambda(XA)| \le 1,\ |1 - \lambda(XA)| = 1\\ \lambda(XA) \ne 0\end{array}\right\} \Rightarrow i(1 - \lambda(XA)) \le k,$

and $i(U_1) \le k$

(xiii) $|\lambda[(I - XA)XA(XA)^d]| \le 1$, $|\lambda[(I - XA)XA(XA)^d]| = 1 \Rightarrow i(\lambda) \le k$,

and $i(U_1) \le k$

(xiv) $C_n^{\,k}[(I - XA)XA(XA)^d]$ converges and $i(U_1) \le k$

(xv) $(I - XA)^n(XA)^d/n^k \to 0$ and $i(U_1) \le k$.

In which case $\lim_n S_n^{\,k} = G =$

$$\begin{cases} X(AX)^\# = (XA)^\# X \quad \text{if } k = 0 \\ \dfrac{(-1)^{k-1}}{k!}\,[I - XA\,(XA)^d](XA)^{k-1}X = \dfrac{(-1)^{k-1}}{k!}\,(XA)^{k-1}X[I - AX(AX)^d] \\ \text{if } k \ge 1. \end{cases} \tag{5.6}$$

<u>Proof</u> We begin by noting that for k = 0, (iv) reduces to $\rho(I - U_1) < 1$, so that U_1^{-1} exists. (i) ⟺ (ii) Obvious as $S_n^{\,k}X^-X = S_n^{\,k}$. (i) ⟺ (iii) Follows from (5.2). The equivalence of (iii), (iv) and (v) follows directly from Theorem (3.8). It should be noted that $\lambda(U_1) = 0 \Rightarrow i(U_1) \le k$, and that this implies with aid of Theorem 4.6, that $i(XA) \le k + 1$.
(v) ⇒ (vi) Clearly $(I - U_1)XX^- = I - U_1$. (vi) ⇒ (vii) Multiply (vi) by

$A(XA)^-$ and use the equivalence of (vi) and (iv). (vii) ⇒ (viii) This result follows by Theorem 3.10 with A replaced by $I - XA$. (viii) ⇒ (ix) Obvious. (ix) ⇒ (v) Apply (3.44) with A replaced by $I - U_1$. Because $i(U_1) \leq k \Rightarrow i(XA) \leq k + 1$, the equivalence of (vii), and (x) through (xv) then follows immediately from (3.43). Before we compute the actual limit, several remarks are in place.

1. We recall that by Theorem 4.6, $i(U_1) \leq k \Leftrightarrow R[(XA)^{k+1}X] = R[(XA)^k X]$

2. By a theorem of Parker [20], $\Delta_{I-U_1}(\lambda) = [\lambda/(\lambda - 1)]^{n-r}\Delta_{I-XA}(\lambda)$, where $r = \text{rank}(X)$, from which the equivalence of (vii) and (viii) may also be seen. Moreover, if $\Delta_{XA}(\lambda) = \lambda^{n_1} \prod_{t=2}^{s} (\lambda - \lambda_t)^{n_t}$, then

$$\Delta_{I-U_1}(\lambda) = \lambda^{n-r} \cdot (\lambda - 1)^{n_1+r-n} \prod_{t=2}^{s} [\lambda - (1 - \lambda_t)]^{n_t}. \tag{5.6a}$$

3. For $k = 0$, we get $n = r = n_1$ and by (4.8) $I - U_1 \approx (I - XA)XA(XA)^{\#}$, which again yields (5.6a). In addition

$$\Delta_{I-U_1^{-1}}(\lambda) = \Delta_{[I-(XA)^{\#}]XA(XA)^{\#}} = \lambda^{n_1} \prod_{2}^{s} [\lambda - (1 - \lambda_t^{-1})]^{n_t}, \tag{5.6b}$$

so that $\rho(I - U_1) = \max_{\lambda(XA)\neq 0} |1 - \lambda(XA)|$, $\rho(I - U_1^{-1}) = \max_{\lambda(XA)\neq 0} |1 - 1/\lambda(XA)|$.

We may now combine Theorem 3.8 and (5.2) to compute $\lim_n S_n^k$. Indeed, we arrive at

$$G = \lim_n S_n^k = \lim_n [C_n^k(I - U_1)]X = \begin{cases} U_1^{-1}X & \text{if } k = 0 \\ \dfrac{(-1)^{k-1}}{k!} U_1^{k-1}(I - U_1U_1^d)X & \text{if } k \geq 1. \end{cases} \tag{5.7}$$

Using the fact from (4.9)xv, that

$$U_1^d = Z^{k+1}(XA)^k XX^- + I - XX^-, \quad U_1^k = (XA)^{k-1}XX^- + I - XX^-$$

we see that for $k = 0$, $G = XYX^-X = XY = ZX = XX^-ZX$, while for $k \geq 1$,
$G = \frac{(-1)^{k-1}}{k!} (I - Z^{k+1}(XA)^{k+1}X)(XA)^{k-1}X = \frac{(-1)^{k-1}}{k!} (I - ZXA)(XA)^{k-1}X$

$= \frac{(-1)^{k-1}}{k!} (XA)^{k-1}X(I - AXY)$. On setting $Y = (AX)^d$ and $Z = (XA)^d$, the result follows. □

From the closed form expression for the limit G, it is now easily verified why the limit has the following properties, of [21],

$$\text{(i)} \quad GA = \begin{cases} (XA)^{\#}XA = P_{R(XA),N(XA)} & \text{if } k = 0 \\ \frac{(-1)^{k-1}}{k!}[\text{Nil}(XA)]^k & \text{if } k \geq 1 \end{cases} \tag{5.8}$$

where $\text{Nil}(M) = M(I - MM^d)$, is the nilpotent part of M.

$$\text{(ii)} \quad AG = \begin{cases} AX(AX)^{\#} = P_{R(AX),N(AX)} & \text{if } k = 0 \\ \frac{(-1)^{k-1}}{k!}[\text{Nil}(AX)]^k & \text{if } k \geq 1 \end{cases}$$

Thus GA and AG are idempotent for $k = 0$ and nilpotent of index 2 for $k \geq 1$.

$$\text{(iii)} \quad GAX = \begin{cases} (XA)^{\#}XAX & \text{if } k = 0 \\ 0 & \text{if } k \geq 1 \end{cases}$$

$$XAG = \begin{cases} XAX(AX)^{\#} & \text{if } k = 0 \\ 0 & \text{if } k \geq 1 \end{cases}$$

because by Theorem 4.6, $X(AX)^d(AX)^{k+1} = (XA)^kX = (XA)^{k+1}(XA)^dX$.

$$\text{(iv)} \quad GAG = \begin{cases} (XA)^{\#}XA(XA)^{\#}X = G & \text{if } k = 0 \\ 0 & \text{if } k \geq 1 \end{cases}.$$

$$\text{(v)} \quad AGA = \begin{cases} A(XA)^{\#}XA & \text{if } k = 0 \\ \frac{(-1)^{k-1}}{k!} A(XA)^k[I - (XA)(XA)^d] & \text{if } k \geq 1 \end{cases}$$

and

$$AGA = A \Leftrightarrow \begin{cases} RS(A) = RS(XA) & \text{if } k = 0 \\ A = 0 & \text{if } k \geq 1 \end{cases}$$

Indeed, for $k \geq 1$,

$$AGA = A \Rightarrow \frac{(-1)^{k-1}}{k!} A(XA)^k[I - XA(XA)^d] = A$$

$$\Rightarrow \frac{(-1)^{k-1}}{k!} (XA)^{k+1}[I - XA(XA)^d] = XA \Rightarrow A + 0 \text{ (by the above).}$$

$$\text{(vi)} \quad G = 0 \Leftrightarrow \begin{cases} X = 0 & \text{if } k = 0 \\ R[(XA)^k X] = R[(XA)^{k-1} X] & \text{if } k \geq 1, \end{cases}$$

because $(XA)^{\#}X = 0 \Rightarrow XA = 0 \Rightarrow X = XAXY = 0 \Rightarrow G = 0$, while for $k \geq 1$, $G = 0 \Leftrightarrow (XA)^k = (XA)^{k+1}(XA)^d \Leftrightarrow i(XA) \leq k$ and $R[(XA)^{k+1}X] = R[(XA)^k X]$ $\Leftrightarrow i(U_1) \leq k - 1$, where we used again Theorem (4.6) (iii), (iv), (viii).

It should be pointed out here, that even though (5.5)viii is equivalent to (5.5)x, the convergence of $(I - XA)^n/n^k$ alone is not equivalent to that of $C_n^k[(I - XA)XA(XA)^-]$. If one selects V_1 instead of U_1 to govern the iteration S_n^k, then one obtains the left analogue of Theorem 5.1. The proof of this follows in analogous fashion, once one recalls the equivalence of (4.9) (vii)-(xii). Similar results may be proven if we select P or Q to govern the iteration S_n^k. The convergence conditions are of course exactly the same, however the computations become more difficult because of the complicated structure of P^d, Q^d. For example, using (4.9)x, it may be shown that

$$\lim_n C_n^k(I - P) = \begin{cases} P^{-1} = (XA)^{\#}XX^- + I - XA(XA)^{\#} & \text{if } k = 0 \\ \frac{(-1)^{k-1}}{k!} P^{k-1}(I - PP^d) = \frac{(-1)^{k-1}}{k!} (XA)^{k-2} \times \\ \quad [I - XA(XA)^d]P & \text{if } k \geq 1, \end{cases} \tag{5.9}$$

which should be compared with (5.6).

We remark that even though the equalities $N(U_1^r) = N(P^r)$ $r \geq 0$, ensure that $Nil(U_1) \approx Nil(P)$, it does not guarantee their equality.

Next, suppose that $(XA)^d$ is known, or more generally that Z in the identity $(XA)^k X = Z^n (XA)^{k+n} X$ $n = 1,2,\ldots$ $k \geq 0$, is known. Then from Theorem 4.6, we know that $I - U_1^d = (I - XA)^d XX^- = [I - Z^{k+1}(XA)^k]XX^-$, and hence that $(I - U_1^d)XX^- = I - U_1^d$. We may now consider the following "inverse" iteration:

$$I_n^k = \sum_{r=0}^{n-1} [I - (XA)^d]^r X/n^k = [C_n^k(I - U_1^d)]X. \tag{5.10}$$

Then $I_n^{\ k}$ converges exactly when $C_n^{\ k}(I - U_1^{\ d})$ does, that is exactly when

$$|1 - \lambda(U_1^{\ d})| \leq 1, \quad |1 - \lambda(U_1^{\ d})| = 1 \Rightarrow i(1 - \lambda(U_1^{\ d}) \leq k.$$

Since the non-zero eigenvalues of U_1 and $U_1^{\ d}$ have the same indices and $i(\lambda_o) = m$ in $\psi_B \Leftrightarrow i(1 - \lambda_o) = m$ in ψ_{1-B}, this reduces to

$$\lambda(U_1) \neq 0 \Rightarrow |1 - 1/\lambda(U_1)| \leq 1, \quad |1 - 1/\lambda(U_1)| = 1 \Rightarrow i(\lambda(U_1)) \leq k; \text{ or}$$

$$\lambda(U_1) \neq 0 \Rightarrow \mathrm{Re}[\lambda(U_1)] \geq \frac{1}{2} \text{ and } \mathrm{Re}[\lambda(U_1)] = \frac{1}{2} \Rightarrow i(\lambda(U_1)) \leq k. \tag{5.11}$$

Recalling that $i(U_1^{\ d}) \leq 1$ and that $i(U_1^{\ d}) = 0 \Leftrightarrow i(U_1) = 0$ and $i(U_1^{\ d}) = 1$ $\Rightarrow i(U_1) \geq 1$, we see that if (5.16) holds, then the actual limit of $I_n^{\ k}$ becomes:

$$\lim_n I_n^{\ k} = \lim_n [C_n^{\ k}(I - U_1^{\ d})] =$$

$$\begin{cases} (U_1^{\ d})^{-1}X = U_1X = XAX & \text{if } k = i(U_1) = 0 \\ 0 & \text{if } k \geq i(U_1) = 0 \\ I - U_1U_1^{\ d} = [I - XA(XA)^d]X & \text{if } k = i(U_1^{\ d}) = 1 \\ 0 & \text{if } k > i(U_1^{\ d}) \geq 1 \end{cases} \tag{5.12}$$

For $k = 0$, (5.16) shows that $I_n^{\ o}$ converges exactly when $\lambda(U_1) \neq 0$ $\Rightarrow \mathrm{Re}[\lambda(U_1)] > \frac{1}{2}$ or equivalently by (5.6b), when $\rho(I - U_1^{\ -1})$ $= \max_{\lambda(XA)\neq 0} |1 - 1/\lambda(XA)| < 1$. We remark that the convergence producing factor X cannot be dropped in the definition of $I_n^{\ k}$, since $C_n^{\ o}(I - (XA)^d)$ cannot converge if XA is singular. The corresponding iterations $C_n^{\ k}(I - P) \cdot X$ and $XC_n^{\ k}(I - Q)$, become tractable only for $k = 0$ in which case we obtain expansions similar to (5.3), (5.4).

$$P = [I - ZXX^-(XA - I)]^{-1} = \sum_{r=0}^{\infty} ZXX^-(XA - I)^r = \lim C_n^{\ o}(I - P^{-1}) \tag{5.13}$$

$$Q = [I - (AX - I)X^-XY]^{-1} = \sum_{r=0}^{\infty} [(AX - I)X^-XY]^r = \lim C_n^{\ o}(I - Q^{-1})$$

where $XAXY = X = ZXAX$. It is easily seen that for $k = 0$ all four iterations have exactly the same convergence conditions. On multiplying (5.4) by XY we obtain:

$$X = \sum_{r=0}^{\infty} (I - Z)^r ZX, \quad \text{Re}[\lambda(XA)] > \frac{1}{2} \quad \text{or} \quad \lambda(XA) = 0 \tag{5.14}$$

or equivalently $\max_{\lambda(XA)\neq 0} |1 - \lambda(XA)| < 1$, which we shall use shortly together with (5.7) in our examples. We now come to the last part of this paper dealing with the "differentiated" iteration T_n^k. The algebraic identities between T_n^k and $D_n^k(I - U_1)$ etc., are step for step analogous to those given in (5.5), (5.9). Consequently, corresponding to each result involving S_n^k, and $C_n^k(I - (\cdot))$ there exists a parallel result for T_n^k and $D_n^k(I - (\cdot))$. The ultimate convergence criterion and limit of course being different, because of the shift in indices encountered in moving from C_n^k to D_n^k. We shall only give one result in detail.

Theorem 5.2 The following are equivalent for $k \geq 0$.

1. T_n^k converges
2. $T_n^k X^-$ converges
3. $D_n^k(I - U_1)$ converges
4. $|1 - \lambda(U_1)| \leq 1$, $|1 - \lambda(U_1)| = 1 \Rightarrow i(1 - \lambda(U_1)) \leq k + 1$, $i(U_1) < k$
5. $(I - U_1)^n/n^{k+1} \to 0$ and $i(U_1) \leq k$
6. $(I - XA)^n X/n^{k+1} \to 0$ and $i(U_1) \leq k$
7. $[(I - XA)XA(XA)^-]^n/n^{k+1} \to 0$ and $i(U_1) \leq k$ (5.15)
8. $(I - XA)^n/n^{k+1}$ converges and $i(XA) \leq k + 1$
9. $(I - XA)^n/n^{k+1} \to 0$
10. $C_n^{k+1}(I - XA)$ converges
11. $|1 - \lambda(XA)| \leq 1$, $|1 - \lambda(XA)| = 1 \Rightarrow i(1 - \lambda(XA)) \leq k + 1$
12. $C_n^k[(I - XA)XA(XA)^d]$ converges
13. $(I - XA)^n(XA)^d/n^{k+1} \to 0$.

In which case, $\lim T_n^k = H =$

$$\begin{cases} (XA)^{\#}X = X(AX)^{\#} & \text{if } k = 0 \\ \dfrac{(-1)^{k-1}}{(k+1)!}[I - XA(XA)^d](XA)^{k-1}X & \text{if } k \geq 1. \end{cases} \tag{5.16}$$

Proof (1) ⟺ (2) Clear as $T_n^k X^- X = T_n^k$. (2) ⟺ (3) This follows from the identities $T_n^k = C_n^k(I - U_1) \cdot X$, $D_n^k(I - U_1) = T_n^k X^- + (I - XX^-)/n^k$. (3) ⟺ (4) Apply (3.48). (4) ⟺ (5) This follows by (3.49) and (3.47) and the fact that $i(U_1) \leq k < k + 1$. (5) ⟺ (6) Clear. (6) ⇒ (7) Clear on multiplying (6) by $X(XA)^-$. (7) ⇒ (8) This follows from the equivalence of (3.43)ii and (iii) and the fact that $i(U_1) \leq k \Rightarrow i(XA) \leq k + 1$. (8) ⇒ (9) By remark (3.44). (9) ⇒ (5) Multiplying (9) by XX^- suffices. The equivalence of (9)-(14) follows directly from (3.37). We remark in passing, that the condition $i(XA) \leq k + 1$ is this time included in the condition $|1 - \lambda(XA)| = 1 \Rightarrow i(1 - \lambda(XA)) \leq k + 1$. Combining (3.48) and (5.7) we finally arrive at $H = \lim_n T_n^k = \lim_n [D_n^k(I - U_1)]X = \frac{1}{k+1} G$, $k \geq 0$, where G is given as in (5.6). The value of H may similarly be computed from $X \lim D_n^k(I - V_1)$, $X \lim D_n^k(I - Q)$ or $\lim [D_n^k(I - P)]X$, under analogous conditions to those given in (5.15). □

We close with several examples illustrating some of the uses of S_n^o.

Examples

1. Outer Inverses $\hat{A}$. If we set $X = \beta\hat{A}$, $\hat{A}A\hat{A} = \hat{A}$ and $\beta > 0$, in S_n^o, then

$$G = \beta\hat{A}(\beta A\hat{A})^{\#} = \hat{A} = \beta \sum_{r=0}^{\infty} (I - \beta\hat{A}A)^r \hat{A}, \tag{5.17}$$

which converges exactly when $\rho[(I - \beta\hat{A}A)\hat{A}(\hat{A})^-] < 1$. Taking $(\hat{A})^- = A$, this reduces to $\rho[(1 - \beta)\hat{A}A] < 1$. Since $\hat{A}A$ is idempotent we arrive at $0 < \beta < 2$, c.f.[1]. Exactly the same expansion is obtained from in (5.12) on setting $X = \hat{B}/\beta$, $A = B$, and $Z = \hat{B}B$, $\beta > 0$. Actually this expansion and the ones derived from it are trivial, since $(1 - \beta\hat{A}A)^k\hat{A} \equiv (1 - \beta)^k\hat{A}$ and $1 \equiv \beta \sum_{r=0}^{\infty} (1 - \beta)^r$ if $0 < \beta < 2$. For outer inverses that possess group inverses, we may use (5.14) with $X = \hat{B}/\beta$, $\beta > 0$, $A = I$, and $Z = \beta(\hat{B})^{\sim}$, where $(\cdot)^{\sim}$ denotes any commutative 1-inverse of $(\cdot)$. Indeed we have

$$\hat{B} = \beta \sum_{r=0}^{\infty} [I - \beta(\hat{B})^{\sim}]^r (\hat{B})^{\sim}\hat{B} = \beta \sum_{r=0}^{\infty} [I - \beta(\hat{B})^{\#}]^r (\hat{B})^{\#}\hat{B}, \tag{5.18}$$

which converges for $\mathrm{Re}[\lambda(\hat{B})] > \frac{1}{2}$ if $\lambda(\hat{B}) \neq 0$.

2. Drazin Inverses A^d. From part (1) we obtain two expansions for A^d, namely

$$A^d = \beta \sum_{r=0}^{\infty} (I - \beta AA^d)^r A^d \qquad 0 < \beta < 2 \tag{5.19}$$

and, because $(B^d)^d = B^2 B^d$,

$$B^d = \beta \sum_{r=0}^{\infty} (I - \beta B)^r BB^d, \tag{5.20}$$

which is valid for $\mathrm{Re}[\lambda(B^d)] > \frac{\beta}{2}$ if $\lambda(B^d) \neq 0$, or equivalently:

$$|\lambda(B) - \frac{1}{\beta}| < \frac{1}{\beta} \quad \text{if } \lambda(B) \neq 0.$$

These expansions are readily verified using canonical forms.

3. Reflexive Inverses A^+. Setting $\hat{A} = A^+$ in (5.22) yields

$$A^+ = \beta \sum_{r=0}^{\infty} (I - \beta A^+ A)^r A^+ \qquad 0 < \beta < 2, \tag{5.21}$$

which also follows from (5.14) on setting $X = B^+/\beta$, $A = B$ and $Z = B^+ B\beta$.

4. Moore-Penrose Inverse $A^\dagger$. Setting $X = \beta A^*$ with $\beta > 0$ in $S_n^{\,o}$ gives

$$G = \beta A^*(AA^*)^{\#} = A^\dagger = \beta \sum_{r=0}^{\infty} (I - \beta A^* A)^r A^*, \tag{5.22}$$

which converges exactly when $\rho[(I - \beta A^* A)A^*(A^*)^-] < 1$. Taking $(A^*)^- = (A^*)^\dagger$, this reduces to $\rho[A^\dagger A - \beta A^* A] < 1$, which is well-known to be equivalent to $0 < \beta < 2\|A\|^{-2}$, [2, p. 298] where $\|A\|$ denotes the spectral norm of A. Similarly if we set in (5.14), $X = B^\dagger/\beta$, $A = B^{*\dagger}$, and $Z = B^* B\beta$, $\beta > 0$, we arrive at exactly the same expansion for $B^\dagger$. This converges for $\mathrm{Re}[\lambda(B^\dagger B^{*\dagger})] = \lambda(B^\dagger B^{*\dagger}) > \beta/2$ if $\lambda(B^\dagger B^{*\dagger}) \neq 0$. Using the singular-value-decomposition of B, this is easily seen to be equivalent to the above condition.

References

1 A. Ben Israel and T. N. E. Greville, Generalized Inverses, Theory and Applications, Wiley, New York, 1974.

2 E. T. Browne, The Theory of Determinants and Matrices, The University of North Carolina Press, Chapel Hill, 1958, 164.

3 R. E. Cline, An application of representations for the generalized inverses of a matrix, MRC Technical Summary Report #592, 1965.

4 C. A. Cullen, Matrices and Linear Transformations, Addison-Wesley, Reading Mass., 1967, first edition, 174.

5 M. P. Drazin, Psuedo-inverses in associative rings and semigroups, Amer. Math. Monthly, 65(1958), 506-514.

6 N. Dunford and J. T. Schwartz, Linear Operators, Vol. 1, Inter. Science Publishers, New York, 1964-71, 560.

7 H. K. Farahat and L. Mirsky, A condition for diagonability of matrices, Amer. Math. Month., 63(1956), 410-412.

8 F. R. Gantmacher, The Theory of Matrices, Vol. 1, Chelsea, New York, 1960.

9 T. N. E. Greville, Spectral generalized inverses of square matrices, MRC Technical Summary Report #832, 1967.

10 R. E. Hartwig and J. Shoaf, On the derivative of the Drazin inverse of a complex matrix, SIAM J. Math. Anal. 10(1979), 207-216.

11 R. E. Hartwig and J. Shoaf, Group inverses and Drazin inverses of bidiagonal and triangular Toeplitz matrices, Austral. J. Math. XXIV (1977), 10-34.

12 R. E. Hartwig, Block generalized inverses, Arch. Rat. Mech., 61(1976), 197-251.

13 R. E. Hartwig, 1-2 inverses and the invariance of BA^+C, Lin. Alg. Appl., 11(1978), 271-275.

14 R. E. Hartwig, Generalized inverses, EP elements and associates, Rev. Roum. Pures/et Appl. XIII (1978), 57-60.

15 R. E. Hartwig and J. Luh, On finite regular rings, Pac. J. Math., 69(1977), 73-95.

16 R. E. Hartwig and J. Luh, A note on unit regular rings, Pac. J. Math., 71(1977), 449-461.

17 G. Losey and H. Schneider, Group membership in rings and semigroups, Pac. J. Math., 11(1961), 1089-1098.

18 J. G. Kemeny and J. L. Snell, Finite Markov Chains, Van Nostrand, New York, 1960.

19 C. D. Meyer, The role of the group generalized inverse in the theory of finite Markov chains, SIAM Rev., Vol. 17, 1975, 443-464.

20 W. V. Parker, On matrices whose characteristic equations are identical, Proc. A.M.S. (1951), 464-466.

21 K. Tanabe, Neumann-type expansion of reflexive generalized inverses of a matrix and the hyperpower iterative method, Lin. Alg. Appl., 19(1975), 163-175.

22 K. Tanabe, An adaptive acceleration of general linear iterative processes for solving a system of linear equations, Proc. 5th Hawaii International Conference on System Science, 1972.

23 J. Von Neumann, On regular rings, Proc. Nat. Acad. Sci., 22(1936), 707-713.

R. E. Hartwig
Mathematics Department
North Carolina State University
Raleigh, North Carolina 27650

and

F. J. Hall
Mathematics Department
Georgia State University
Atlanta, Georgia 30303

E CHANG

The generalized inverse and interpolation theory

1. INTRODUCTION

The Moore-Penrose theory provides a generalized inverse for every matrix, square or rectangular, having some properties of the usual inverse. For a particular class of operators between Hilbert spaces, a Moore-Penrose generalized inverse can be defined (Section 2). Interpolation theory provides an interesting application for the generalized inverse.

The prototype interpolation problem considered is as follows: Given the class of functions X, and linear functional interpolation conditions with associated data, find the element in X that best satisfies the conditions in an extremal sense to be specified.

In Section 3, the class of interpolants X is selected as P_n, polynomials of degree n or less. H^2, the classical Hardy space of analytic functions, is the underlying space for Sections 4 and 5. The problem of a finite number of interpolation conditions for finite point sets $\{(x_i, y_i)\}_1^n$ (Section 4) was studied by Satoru Takenaka [10] and his minimum norm interpolant will be shown to be a generalized inverse solution. The extension to an infinite set of interpolation conditions (Section 5) was considered by Satoru Takenaka [10] and J. L. Walsh [11]. These solutions, derived using the classical theory of analytic functions, are also shown to be generalized inverse solutions.

The theory of the generalized inverse provides a new approach to interpolation theory and an updated interpretation.

2. GENERALIZED INVERSES

The operators considered here will be bounded, linear mappings with closed ranges between Hilbert spaces H_1 and H_2. Let T be such an operator: $T: H_1 \to H_2$.

The following notation will be employed:

$T^\dagger$ - the generalized inverse of T

T^* - the adjoint of T

$N(T)$ - the null space of T

$R(T)$ - the range of T

$M^{\perp}$ - the orthogonal complement of the subspace M

P_M - the projection operator onto the subspace M

Recall that $R(T)^{\perp} = N(T^*)$ and $N(T)^{\perp} = \overline{R(T^*)}$.

It is well known that every bounded linear operator $T: H_1 \to H_2$ with closed range possesses a unique generalized inverse (GI), $T^{\dagger}$, characterized as the unique linear operator X which satisfies any of the following sets of equations:

$$\text{(i)} \quad XTX = X,\ (XT)^* = XT,\ TXT = T,\ (TX)^* = TX \tag{2.1}$$

$$\text{(ii)} \quad XTx = x \text{ for all } x \text{ in } R(T^*) \text{ and } Xy = 0 \text{ for all } y \text{ in } N(T^*) \tag{2.2}$$

$$\text{(iii)} \quad TX = P_{R(T)},\ XT = P_{R(X)}\ . \tag{2.3}$$

Equations (2.1) are the Penrose equations; equations (2.3) are the Moore equations; equations (2.2) are the function-theoretic formulation of $T^{\dagger}$. Equations (2.2) may also be viewed as a geometric formulation. The unique GI will be termed the Moore-Penrose GI due to (2.1) and (2.3). For a full development of these properties and other equivalent characterizations, see [5], [7], [8].

As shown in [7], the GI has some important implications in the solution of the linear operator equation:

$$Tx = y. \tag{2.4}$$

The element $x^o = T^{\dagger}y$, to be referred to as the GI solution of equation (2.4), satisfies the following:

$$||Tx^o - y|| \leq ||Tx - y|| \quad \text{for all } x \in H_1 \tag{2.5}$$

$$||x^o|| \leq ||x|| \text{ for all } x \text{ producing equality in (2.5).} \tag{2.6}$$

In other words, the GI solution is the minimum norm least squares solution to (2.4). The least squares property (2.5) is equivalent to solving the normal equations

$$T^*Tx = T^*y\ . \tag{2.7}$$

Furthermore, the following representation of the GI of T is valid,

$$T^{\dagger} = (T^*T)^{\dagger}T^*\ . \tag{2.8}$$

3. POLYNOMIAL INTERPOLATION

The polynomial $p_n \in P_n$ that passes through $n+1$ specified values at $n+1$ abscissae is well known to be unique if the abscissae are distinct. This is directly related to the nonsingularity of the Vandermonde matrix (V in (3.2)). The pointwise interpolation conditions are

$$p_n(z_k) = \sum_{i=0}^{n} a_i z_k^i = y_k \quad , \quad k=0,1,\ldots,n \tag{3.1}$$

where $\{a_i\}_{i=0}^n$ are the coefficients of $p_n(z)$, $\{y_k\}_{k=0}^n$ are the interpolation data and $\{z_k\}_{k=0}^n$ are the distinct abscissae. Equations (3.1) are equivalent to the <u>interpolation system</u>

$$\begin{bmatrix} 1 & z_o & \cdots & z_o^n \\ \vdots & & & \vdots \\ 1 & z_n & \cdots & z_n^n \end{bmatrix} \begin{bmatrix} a_o \\ \vdots \\ a_n \end{bmatrix} = \begin{bmatrix} y_o \\ \vdots \\ y_n \end{bmatrix}$$

$$VA = Y. \tag{3.2}$$

In this section we consider sets of linear functional interpolation conditions, $L_k p_n(z) = y_k$, $k=0,1,\ldots,m$ in which the resulting interpolation system

$$\begin{bmatrix} L_o z^o & \cdots & L_o z^n \\ \vdots & & \vdots \\ L_m z^o & \cdots & L_m z^n \end{bmatrix} \begin{bmatrix} a_o \\ \vdots \\ a_n \end{bmatrix} = \begin{bmatrix} y_o \\ \vdots \\ y_m \end{bmatrix}$$

represented as

$$DA = Y \tag{3.3}$$

is solved using the GI.

The polynomial whose coefficients are the components of the GI solution of (3.3), $A^o := D^\dagger Y$, will be called the <u>GI interpolating polynomial.</u>

There are two useful representations for the unique pointwise interpolating polynomial. The <u>Lagrange representation</u> for $p_n(z)$ satisfying (3.1) is

$$p_n(z) = \sum_{k=0}^{n} y_k \frac{\omega_n(z)}{(z-z_k)\,\omega'(z_k)} \tag{3.4}$$

$$\omega_n(z) = \prod_{j=0}^{n} (z-z_j). \qquad (3.5)$$

For a fixed set of distinct abscissae and data, the Lagrange Representation is easy to determine and convenient to use. However, to proceed to a higher degree polynomial an entirely new calculation is needed. The Newton representation has the permanence property, that only the addition of terms to the previous interpolant is required. The unique polynomial satisfying (3.1) has the Newton representation:

$$p_n(z) = \sum_{i=0}^{n} c_i\omega_i(z) \quad . \qquad (3.6)$$

The coefficients c_i are determined by successively setting $z = z_o$, $z = z_1, \ldots,$ and solving the resulting equations. The permanence property was attained at the expense of the added complexity of the multipliers c_i.

As shown in Section 5, the Lagrange and Newton representations play a vital role in the transition from finite to infinite interpolation by analytic functions.

Let X be a linear space. Given a linear functional L and values $y_o, y_1, \ldots, y_m$, the term multidata will refer to the following kind of interpolation conditions:

$$Lp = y_k,\ k=0,\ 1,\ \ldots,\ m,\ p \in X \qquad (3.7)$$

The problem of polynomial interpolation to multidata (3.7) induces the interpolation system

$$\begin{bmatrix} 1 & z_o & z_o^2 & \ldots & z_o^n \\ \vdots & & \vdots & & \\ 1 & z_o & z_o^2 & \ldots & z_o^n \end{bmatrix} \begin{bmatrix} a_o \\ a_1 \\ \vdots \\ a_n \end{bmatrix} = \begin{bmatrix} y_o \\ \vdots \\ y_m \end{bmatrix}.$$

$$V_{mn}A_n = Y_m \qquad (3.8)$$

The GI interpolating polynomial can be written

$$\tilde{p}_n(z) = V_n(z) \cdot \tilde{A}_n \qquad (3.9)$$

where $\tilde{A}_n = V^{\dagger}_{mn} Y_m$ is the GI solution of (3.8) and

$$V_n(z) = [1\ z\ z^2\ .\ .\ .\ z^n]\ . \qquad (3.10)$$

Recalling the inconsistency of the problem, the immediate question is what value is returned by the GI polynomial. Since the GI solution $\tilde{A}_n$ satisfies the normal equations(2.7),

$$V^*_{mn}(V_{mn}\tilde{A}_n - Y_m) = 0 \quad , \tag{3.11}$$

or equivalently

$$\overline{\begin{bmatrix} 1 & \cdots & 1 \\ z_o & \cdots & z_o \\ \vdots & & \vdots \\ z_o^h & & z_o^n \end{bmatrix}} \begin{bmatrix} V_n(z_o) \cdot \tilde{A}_n - y_o \\ \vdots \\ V_n(z_o) \cdot \tilde{A}_n - y_m \end{bmatrix} = \begin{bmatrix} 0 \\ 0 \\ \vdots \\ 0 \end{bmatrix} , \tag{3.12}$$

it follows that $\sum_{k=0}^{m} (\tilde{p}_n(z_o) - y_k) = 0$, or

$$p_n(z_o) = \frac{1}{m+1} \sum_{k=0}^{m} y_k \quad .$$

The GI approach selects the element in $\mathcal{P}_n$ which passes through the multidata average, and has the minimal sum of the squares of the coefficients. This averaging ability generalizes to the case of a finite number of multidata conditions (Chang [2]). For the general case of <u>any</u> finite set of linear functional conditions, $\{L_i\}^m_{k=0}$ we turn from an algebraic to an analytic approach.

The problem now is to find $p_n \in \mathcal{P}_n$ such that

$$L_k p_n = y_k \quad , \quad k=0,1,\ldots,m \quad . \tag{3.14}$$

Note that the degree of the polynomial and the number of conditions are unrestricted.

The associated interpolation system is

$$\begin{bmatrix} L_o 1 & L_o z & \cdots & L_o z^n \\ \vdots & & & \\ \vdots & & & \\ L_m 1 & L_m z & \cdots & L_m z^n \end{bmatrix} \begin{bmatrix} a_o \\ \vdots \\ \vdots \\ a_n \end{bmatrix} = \begin{bmatrix} y_o \\ \vdots \\ y_m \end{bmatrix} ,$$

$$DA_n = Y_m \quad . \tag{3.15}$$

Theorem 1. Let $p_n(x)$ be a polynomial whose vector of coefficients is $\tilde{A}_n = D^{\dagger}Y_m$ with (i) $Y_m \in R(D)$, a consistent problem, or (ii) $Y_m \notin R(D)$, an inconsistent problem. Then $p_n(z)$ is either (i) the interpolating polynomial or (ii) the least squares polynomial approximation, such that the Euclidean norm $||\tilde{A}_n||$ is a minimum.

Proof It follows directly from the minimum norm least squares properties (2.5-2.6) of the GI solution. □

From these simple interpolation problems the GI interpretation provides an efficient way of dealing with inconsistent problems.

4. FINITE INTERPOLATION IN A HARDY SPACE

In generalizing to interpolation by analytic functions, the minimum norm property of the GI approach to polynomial interpolation is retained if the Hardy space, H^2, is considered as the class of interpolants. The choice is obvious upon examination of the norm and associated inner product.

The finite interpolation problem is to find $f \in H^2$ such that

$$f(z_k) = y_k, \quad k=0,1,\ldots,m \tag{4.1}$$

where $\{z_k\}_{k=0}^m$ are distinct, $|z_i| < 1$ and $\{y_k\}_{k=0}^m$ are the given interpolation data.

This problem was studied by Satoru Takenaka [10], in the 1920's. The concept of the GI provides a new approach to this classical problem. Takenaka's unique minimum norm interpolant is shown here to be the function attained through the theory of generalized inverses.

The *Hardy space* H^2 is the set of single-valued analytic functions f defined on the unit ball $B_1 = \{z : |z|<1\}$ for which

$$||f|| = \lim_{r\to 1} \left(\frac{1}{2\pi} \int_{|z|=r} |f(z)|^2 |dz|\right)^{1/2} \tag{4.2}$$

is finite. The associated inner product is

$$(f,g) = \lim_{r\to 1} \frac{1}{2\pi} \int_{|z|=r} f(z)\overline{g(z)}\,|dz| \quad . \tag{4.3}$$

Some attributes of H^2 are listed below (see Rudin [9]):

(i) $\{z^k\}_{k=0}^\infty$ forms a complete orthonormal system.

(ii) The function $f(z) = \sum_{k=0}^\infty a_k z^k$ is an element of H^2 if and only if $\sum_{k=0}^\infty |a_k|^2 < \infty$.

(iii) The inner product of $f(z) = \sum_{k=0}^\infty a_k z^k$ and $g(z) = \sum_{k=0}^\infty b_k z^k$ is

$$(f,g) = \sum_{k=0}^{\infty} a_k \overline{b_k} \quad . \tag{4.4}$$

(iv) For $w \in B_1$, and any $f \in H^2$, the point evaluation linear functional $Lf = f(w)$ is bounded for all w in a closed subset of B_1, with

$$|f(w)| \leq R^{-1/2} \| f \|, \quad \text{where } R: = r-|w| > 0. \tag{4.5}$$

As a consequence of (iv), H^2 is a Hilbert space admitting a reproducing kernel function.

Let S be a point set in $\mathbb{C}$, the field of complex numbers, and X a separable Hilbert space of functions defined on S. A function $K(P,Q)$ of two variables P and Q in S is called a <u>reproducing kernel function</u> (RK) for the space X if (i) for each fixed $Q \in S$, $K(P,Q)$, as a function of P, is in X and (ii) for any $f(P) \in X$ and for every Q in S the reproducing property

$$f(Q) = (f(P),K(P,Q)) \tag{4.6}$$

holds, where the inner product is formed on the P variable (Davis [3]).

Some relevant properties of the RK are listed below (Davis [3]):

(i) If a RK exists, it is unique

(ii) $K(P,Q) = \overline{K(Q,P)}$

(iii) The representer of L, a bounded linear functional defined on X, is

$$L(Q) = \overline{L_p K(P,Q)} \quad , \tag{4.7}$$

that is, the conjugate of L applied to $K(P,Q)$ as a function of P.

Let $\{\Phi_k\}_{k=0}^{\infty}$ be a complete orthonormal system for H^2 with $z,t \in B_1$. The unique RK is obtained by forming the series (Hille [6]) $K(z,t) = \sum_{k=0}^{\infty} \Phi_k(z) \overline{\Phi_k(t)}$. Since $\{z^k\}_{k=0}^{\infty}$ is such a system, the representation of the RK for H^2 is particularly simple:

$$K(z,t) = \sum_{k=0}^{\infty} z^k \bar{t}^k = (1-z\bar{t})^{-1} \text{ for all } z,t \in B_1 \tag{4.8}$$

Associated with the RK is a minimum norm property evidenced in the one-point interpolation problem of finding $f \in H^2$ such that for fixed $t \in B_1$,

$$\begin{cases} f(t) = 1 \\ \|f\| \text{ is a minimum.} \end{cases} \tag{4.9}$$

The function $f(z) = \dfrac{K(z,t)}{K(t,t)}$ is the unique minimum norm interpolant (Bergman [1]).

Theorem 2. The functions $\{K(z,z_i)\}_{i=0}^{\infty}$ are linearly independent in H^2 if $\{z_i\}_{i=0}^{\infty}$ are distinct in B_1.

Proof For any integer $N > 0$, form the linear combination $\sum_{j=0}^{N} c_j K(z,z_j) = 0$. For any $f \in H^2$, $0 = (f(z),0) =$

$(f(z), \sum_{j=0}^{N} c_j K(z,z_j)) = \sum_{j=0}^{N} \overline{c_j}\, f(z_j)$ by the reproducing property (4.6).

Since this relation must hold for all $f \in H^2$ with N+1 distinct abscissae, then $c_i=0$, $i = 0,1,\ldots,N$, with N arbitrary. Thus the $\{K(z,z_i)\}_{i=0}^{\infty}$ are linearly independent. □

Takenaka [10] noted that the general solution of (4.1) consists of the sum of a homogeneous solution f_o which satisfies

$$f_o(z_k) = 0, \quad k=0,1,\ldots,m \qquad (4.10)$$

and a particular solution which is the linear combination of the $\{K(z,z_j)\}_{j=0}^{m}$ that satisfies (4.1). Takenaka proved that the particular solution is the unique minimum norm interpolant and is representable using Cramer's rule as

$$f(z) = \frac{\sum_{k=0}^{m} y_k \begin{vmatrix} K(z_o,z_o) & \ldots & K(z_o,z_m) \\ \vdots & & \vdots \\ K(z,z_o) & \ldots & K(z,z_m) \\ \vdots & & \\ K(z_m,z_o) & \ldots & K(z_n,z_m) \end{vmatrix} \leftarrow k+1^{th} \text{ row}}{\begin{vmatrix} K(z_o,z_o) & \ldots & K(z_o,z_m) \\ \vdots & & \vdots \\ K(z_m,z_o) & \ldots & K(z_m z_m) \end{vmatrix}} \qquad (4.11)$$

or equivalently

$$f(z) = \sum_{j=0}^{m} y_j \frac{G_m^j(z)}{G_m} \qquad (4.12)$$

where G_m is the Gram determinant of $\{K(z,z_i)\}_{i=0}^{m}$ and G_m^j is the Gram determinant with the $j+1^{th}$ row replaced by $\{K(z,z_i)\}_{i=0}^{m}$.

Substituting the expression for the RK of H^2 and simplifying in a manner similar to the evaluation of the Cauchy determinant (Davis [3]) results in an expression in terms of the abscissae and data

$$f(z) = \prod_{k=0}^{m} y_k \frac{1}{z-z_k} \prod_{j=0}^{m} \frac{z-z_j}{1-z\bar{z}_j} \frac{\prod_{j=0}^{m} 1-z_k\bar{z}_j}{\prod_{\substack{j=0 \\ j\neq k}}^{m} z_k - z_j} . \tag{4.13}$$

Takenaka defined the following function B_n that we recognize as the <u>finite Blaschke product</u> for the abscissae $\{z_i\}_{i=0}^{m}$

$$B_m(z) = \prod_{j=0}^{m} \frac{z-z_j}{1-z\bar{z}_j} . \tag{4.14}$$

Takenaka's minimum norm interpolating function can be expressed in terms of the finite Blaschke product

$$f(z) = \prod_{k=0}^{m} y_k \frac{B_m(z)}{(z-z_k)B_m'(z_k)} , \text{ with } B_m'(z_k) = \frac{\prod_{j=0,\neq k}^{m} z_k - z_j}{\prod_{j=0}^{m} 1-z_k\bar{z}_j} \tag{4.15}$$

The minimum norm property of Takenaka's solution makes this interpolation problem (4.1) attractive for a generalized inverse interpretation.

Given a finite set of distinct abscissae $\{z_i\}_{i=0}^{m}$ in B_1, we define a linear operator Λ_m on H^2 into $\mathbb{C}^{m+1}$, the m+1 unitary space, by

$$\Lambda_m f = \{f(z_i)\}_{i=0}^{m} \tag{4.16}$$

Λ_m <u>will be called the finite interpolation operator</u>.

<u>Theorem 3.</u> The finite interpolation operator Λ_m has the following properties:

(i) $D(\Lambda_m) = H^2$

(ii) $R(\Lambda_m) = \mathbb{C}^{m+1}$

(iii) Λ_m is a bounded operator.

<u>Proof</u> (4.15) exhibits one function that will interpolate any set of values specified at m+1 distinct abscissae, thus proving (ii): The norm (Euclidean) of $\Lambda_m f$ is $||\Lambda_m f|| = \left(\sum_{i=0}^{m} |f(z_i)|^2\right)^{1/2}$. From the inequality

(4.5) we have that $|f(z_i)| \leq R^{-1/2}||f||$, $i=0,1,\ldots,m$, where $R = 1 - \max_{0 \leq i \leq m} |z_i| > 0$. Therefore $||\Lambda_m f|| \leq (\frac{m+1}{R})^{1/2}||f||$ for all $f \in H^2$, and (iii) is proved. □

As a consequence of these properties, the bounded linear operator, Λ_m, between Hilbert spaces, with $R(\Lambda_m)$ closed, has a Moore-Penrose generalized inverse. Given a set of interpolation data $Y_m \in ¢^{m+1}$, the following inequalities are associated with the GI of Λ_m (see (2.5) and (2.6)):

$$||\Lambda_m \Lambda_m^{\dagger} Y_m - Y_m|| \leq ||\Lambda_m f - Y_m|| \quad \text{for all } f \in H^2 \tag{4.17}$$

$$||\Lambda_m^{\dagger} Y_m|| < ||f|| \text{ for all } f \neq \Lambda_m^{\dagger} Y_m \tag{4.18}$$

producing equality in (4.17).

Takenaka's solution (4.15) to the finite interpolation problem can be interpreted in terms of an operator mapping a vector in $¢^{m+1}$ into its minimum norm interpolating function in H^2.

By inspection, the operator, called the Takenaka solution operator,

$$X : ¢^{m+1} \to H^2$$

$$X = \sum_{k=0}^{m} \frac{B_m(z)}{(z-z_k)B'_m(z_k)} e_k \, , \tag{4.19}$$

where $\{e_k\}_{k=0}^{m}$ is the standard basis of $¢^{m+1}$ (e_k is the vector having 1 in the $k+1^{st}$ position and zeros elsewhere), maps any set of interpolation data $Y_m \in ¢^{m+1}$ into the Takenaka solution (4.15) to the finite interpolation problem (4.1).

The Takenaka solution operator X together with the finite interpolation operator Λ_m possesses certain projection properties which will be used to prove the following theorem.

Theorem 4. The Takenaka solution operator is the Moore-Penrose GI of the finite interpolation operator Λ_m.

Proof The null space of Λ_m is the set of homogeneous interpolating functions (4.10)

$$N(\Lambda_m) = \{f_o \in H^2 : \Lambda_m f_o = 0\} \quad . \tag{4.20}$$

It is easy to verify that $N(\Lambda_m)$ is orthogonal to the span of $\{K(z,z_i)\}_{i=0}^m$. Thus

$$N^{\perp}(\Lambda_m) = \{f \in H^2 : f = \sum_{i=0}^{m} c_i K(z,z_i)\} \tag{4.21}$$

But the Takenaka solution is in $N^{\perp}(\Lambda_m)$ and hence

$$R(X) = N^{\perp}(\Lambda_m) \tag{4.22}$$

To prove that X is the GI of Λ_m, we will prove that X satisfies the projection properties (2.3).

For any $Y_m \in \mathbb{C}^{m+1}$, $\Lambda_m X Y_m = \Lambda_m F = Y_m$ where $F \in N^{\perp}(\Lambda_m)$. Thus we have $\Lambda_m X = I$, the identity operator on $\mathbb{C}^{m+1}$. For any $f \in H^2$, $X\Lambda_m f =$ $X\{f(z_i)\}_{i=0}^m = F(z) \in N^{\perp}(\Lambda)$ such that $F(z_i) = f(z_i)$, $i=0,1,\ldots,m$. So $(X\Lambda_m)^2 f = X\{F(z_i)\}_{i=0}^m = F(z)$. Therefore $X\Lambda_m$ is idempotent.

For any f, $g \in H^2$, we consider the inner product $(X\Lambda_m f,g)$. Using the orthogonal decomposition induced by the subspace $N(\Lambda_m)$, the inner product becomes

$$(X\Lambda_m P_{N(\Lambda_m)} f, P_{N(\Lambda_m)} g) + (X\Lambda_m P_{N^{\perp}(\Lambda_m)} f, P_{N(\Lambda_m)} g)$$

$$+ (X\Lambda_m P_{N(\Lambda_m)} f, P_{N^{\perp}(\Lambda_m)} g) + (X\Lambda_m P_{N^{\perp}(\Lambda_m)} f, P_{N^{\perp}(\Lambda_m)} g).$$

The first, second and third terms are zero and in the last term $X\Lambda_m P_{N^{\perp}(\Lambda_m)} f = P_{N^{\perp}(\Lambda_m)} f$ and $P_{N^{\perp}(\Lambda_m)} g = X\Lambda_m P_{N^{\perp}(\Lambda_m)} g$. Thus

$(X\Lambda_m f, g) = (P_{N^{\perp}(\Lambda_m)} f, X\Lambda_m P_{N^{\perp}(\Lambda_m)} g) + (P_{N(\Lambda_m)} f, X\Lambda_m P_{N^{\perp}(\Lambda_m)} g)$

$+ (P_{N^{\perp}(\Lambda_m)} f, X\Lambda_m P_{N(\Lambda_m)} g) + (P_{N(\Lambda_m)} f, X\Lambda_m P_{N(\Lambda_m)} g)$. The last three terms being zero, we get $(X\Lambda_m f, g) = (f, X\Lambda_m g)$. Hence $X\Lambda_m$ is Hermitian and $X\Lambda_m$ is a projection onto the subspace $R(X)$, thus $X\Lambda_m = P_{R(X)}$. By the characterization (2.3), X is the GI of Λ_m and will hereafter be denoted by

$$\Lambda_m^{\dagger} = \sum_{k=0}^{m} \frac{B_m(z)}{(z-z_k)B_m'(z_k)} e_k \quad . \tag{4.23}$$

This theorem shows that Takenaka's solution to the finite interpolation problem is indeed the GI solution. The concept of the generalized inverse enables us to rephrase and update Takenaka's work.

The Associated Vandermonde Operator, V_m

The interpolation conditions (3.15) of the polynomial interpolation problem of Section 3 where $\{L_k\}_{k=0}^{m}$ are the pointwise linear functionals for the abscissae $\{z_k\}_{k=1}^{n}$ produces the associated interpolation system

$$V_{mn}A_n = Y_m \tag{4.24}$$

where V_{mn} is the Vandermonde matrix $[[z_i^j]]_{i=0,1,\ldots,m,j=0,1,\ldots n}$. The interpolating polynomial with minimum sum of the squares of the coefficients is given in terms of its coefficient vector $A_n = V_{mn}^{\dagger} Y_m$, the GI solution of (4.24). In other words, the interpolation problem in $\mathcal{P}_n$ was solved as a linear system in the coefficient space $\mathbb{C}^{n+1}$.

The finite interpolation operator Λ_m is a mapping from an infinite dimensional function space H^2 into $\mathbb{C}^{m+1}$. In a similar manner the interpolation system (4.24) can be solved to determine the coefficients of the Takenaka solution.

The operator, V_m, mapping ℓ^2, the Hilbert space of square summable sequences, into $\mathbb{C}^{m+1}$ defined by

$$V_m A = \begin{bmatrix} 1 & z_o & z_o^2 & \cdots \\ \vdots & & & \\ 1 & z_m & z_m^2 & \cdots \end{bmatrix} \begin{bmatrix} a_o \\ a_1 \\ \vdots \end{bmatrix} = \begin{bmatrix} y_o \\ \vdots \\ y_m \end{bmatrix} \tag{4.25}$$

will be called the Vandermonde operator of dimension $(m+1)\times\infty$, for the distinct abscissae $\{z_i\}_{i=0}^{m}$. In terms of V_m

$$\begin{cases} V_m A = Y_m \\ ||A||_{\ell^2} \text{ is a minimum} \end{cases} \tag{4.26}$$

is the finite interpolation problem posed in ℓ^2, the coefficient space.

Theorem 5. The following is an expression for $V_m^{\dagger}$ in terms of the Blaschke product (4.14)

$$V_m^{\dagger} = \left[\left[\left(\frac{B_m(z)}{(z-z_j)B_m'(z_j)}, z^i\right)\right]\right] \qquad \begin{matrix} i=0,1,\ldots \\ j=0,1,\ldots,m \end{matrix} \tag{4.27}$$

Proof The vector of Fourier coefficients of the Takenaka solution is the GI solution of (4.26) by construction.

$$A = V_m^{\dagger} Y_m = \left\{\left(\Lambda_m^{\dagger} Y_m, z^k\right)\right\}_{k=0}^{\infty} . \tag{4.28}$$

The representation (4.27) for $V_m^{\dagger}$ is obtained by substituting the Takenaka solution operator (4.19) into (4.28). □

Note that the k+1$^{\underline{st}}$ column of $V_m^{\dagger}$ is the coefficient vector of

$\dfrac{B_m(z)}{(z-z_k)B_m'(z_k)}$ expanded in powers of z.

We now consider further the algebraic connection between $\Lambda_m^{\dagger}$ and $V_m^{\dagger}$. $\Lambda_m^{\dagger}$ will be expressed solely in terms of the abscissae $\{z_i\}_{i=0}^{m}$ and interpolation data $\{y_i\}_{i=0}^{m}$.

Notation: The Vandermonde operator can be expressed in terms of the vectors $v_k = (1 \quad z_k \quad z_k^2 \ \ldots)$, $k=0,1,\ldots,m$, $V_m = \begin{bmatrix} v_o \\ \vdots \\ v_m \end{bmatrix}$. $V(z)$ will be used to denote the vector of basis functions $V(z) = (1 \quad z \quad z^2 \quad \ldots)$.

Theorem 6. $\Lambda_m^{\dagger}$ and V_m satisfy the following relations

$$\text{(i)} \quad \Lambda_m^{\dagger} = V(z)\cdot V_m^*(V_m\cdot V_m^*)^{-1} \tag{4.29}$$

$$\text{(ii)} \quad V_m^{\dagger} = V_m^*(V_m\cdot V_m^*)^{-1} \tag{4.30}$$

$$\text{(iii)} \quad \Lambda_m^{\dagger} = V(z)\cdot V_m^{\dagger} \tag{4.31}$$

Proof From (4.12) an equivalent form for the Takenaka solution operator in terms of the Gram determinant is

$$\Lambda_m^{\dagger} = \sum_{k=0}^{m} y_k \frac{G_m^k(z)}{G_m} e_k .$$

$K(z_i,z_j)$ can be written in terms of v_j, $j=0,1,\ldots,m$

$$K(z_i,z_j) = \sum_{k=0}^{\infty} z_i^k \bar{z}_j^k = v_i \cdot v_j^* \text{ , so that,}$$

$$G_m = \begin{vmatrix} v_o \cdot v_o^* & v_o \cdot v_1^* & \cdots & v_o \cdot v_m^* \\ \vdots & & & \\ v_m \cdot v_o^* & v_m \cdot v_1^* & \cdots & v_m \cdot v_m^* \end{vmatrix} = |V_m \cdot V_m^*| \text{ .}$$

In a similar manner $K(z,z_i) = V(z)\cdot v_i^*$

$$G_m^k(z) = \begin{vmatrix} v_o \cdot v_o^* & \cdots & v_o \cdot v_m^* \\ \vdots & & \\ V(z)\cdot v_o^* & \cdots & V(z)\cdot v_m^* \\ \vdots & & \\ v_m \cdot v_o^* & \cdots & v_m \cdot v_m^* \end{vmatrix} \quad \leftarrow \text{ k+1}\underline{\text{st}} \text{ row.}$$

Expanding $G_m^k(z)$ about the k+1st row and denoting the determinant of the minor of the ki<u>th</u> element by M_{ki} produces

$$G_m^k(z) = V(z)\cdot \Big[(-1)^k v_o^* \cdot M_{ko} + (-1)^{k+1} v_1^* \cdot M_{k1} + \ldots + (-1)^{k+m} v_m^* \cdot M_{km}\Big]$$

$$\sum_{k=0}^{m} G_m^k(z) e_k = V(z) \sum_{k=0}^{m} \Big[(-1)^k v_o^* M_{ko} + (-1)^{k+1} v_1^* M_{k1} + \ldots + (-1)^{k+m} v_m^* M_{km}\Big] e_k$$

$$= V(z)[v_o^*, v_1^* \ldots v_m^*] \begin{bmatrix} (-1)^o M_{oo} & \cdots & (-1)^m M_{mo} \\ \vdots & & \\ (-1)^m M_{om} & \cdots & (-1)^{2m} M_{mm} \end{bmatrix}$$

$$= V(z)\cdot V_m^* \cdot \text{adj}[V_m \cdot V_m^*]$$

where the adjoint of a matrix B, adj B, is defined to be the transpose of the matrix obtained from B by replacing each element by its cofactor. Thus

$$\Lambda_m^{\dagger} = V(z) \cdot |V_m \cdot V_m^*|^{-1} V_m^* \cdot \text{adj}(V_m \cdot V_m^*) \text{ .}$$

We note that if a matrix B is nonsingular, then $B^{-1} = (\det B)^{-1}\text{adj } B$. The distinct abscissae guarantee that $V_m \cdot V_m^*$ is nonsingular. Hence $\Lambda_m^\dagger = V(z) \cdot V_m^*(V_m \cdot V_m^*)^{-1}$. We observe that by design the GI solution of (4.26) is the coefficient vector of the Takenaka solution. Therefore we have the identification of operators $\Lambda_m^\dagger = V(z) \cdot V_m^\dagger$ and as a consequence $V_m^\dagger = V_m^*(V_m \cdot V_m^*)^{-1}$. □

The expression for $V_m^\dagger$, (4.30), is a generalization of the least squares formulation of the GI given by (2.8).

5. THE INFINITE INTERPOLATION PROBLEM IN A HARDY SPACE

Let $\{z_i\}_{i=0}^\infty$ be a fixed sequence of distinct complex numbers in B_1. Given a set of interpolation data $Y = \{y_i\}_{i=0}^\infty \in \mathbb{C}^\infty$, we pose the infinite interpolation problem to find $F \in H^2$, if it exists, such that

$$F(z_i) = y_i \ , \ i=0,1,\ldots, \tag{5.1}$$
$$||F|| \text{ is minimum.}$$

Takenaka [10] and Walsh [11] considered this problem and determined necessary and sufficient conditions for the existence and uniqueness of the interpolating function of minimum norm. The following discussion is Takenaka's and Walsh's derivation.

The Takenaka solution (4.15) of the finite interpolation problem provides a starting point. This representation is a Lagrange representation and is convenient for studying the feasibility of the minimum norm solution being given as

$$F(z) = \lim_{m\to\infty} \sum_{k=0}^{m} y_k \frac{B_m(z)}{(z-z_k)B'_m(z_k)} \ , \tag{5.2}$$

the limit of the finite interpolant for the abscissae

$$\{z_i\}_{i=0}^m \subset \{z_i\}_{i=0}^\infty \ .$$

For distinct abscissae, $\{z_i\}_{i=0}^\infty$, the set of functions $\{K(z,z_i)\}_{i=0}^\infty$ was shown to be linearly independent. The functions, denoted by $\{T_k\}_{k=0}^\infty$, formed by the Gram-Schmidt process to orthonormalize $\{K(z,z_i)\}_{i=0}^\infty$ will be called the Takenaka-Walsh orthonormal system (Takenaka [10] p. 143):

$$T_k(z) = \frac{\sqrt{1-|z_k|^2}\ (z-z_o)\ \ldots\ (z-z_{k-1})}{(1-z\bar{z}_o)\ \ldots\ (1-z\bar{z}_k)}\ , \ k=0,1,\ldots \tag{5.3}$$

The equivalent Newton representation of the minimum norm finite interpolant for the abscissae $\{z_i\}^m_{i=0}$ is

$$F_m(z) = \sum_{k=0}^{m} \frac{y_k - F_{k-1}(z_k)}{\sqrt{1-|z_k|^2}|B_k'(z_k)|} T_k(z) \quad . \tag{5.4}$$

With this representation Takenaka [10] and Walsh [11] easily established a necessary and sufficient condition for a solution to the infinite interpolation problem (5.1) to exist and be given by (5.2), the Lagrange form or equivalently, by the Newton form

$$F(z) = \lim_{m\to\infty} \sum_{k=0}^{m} \frac{y_k-F_{k-1}(z_k)}{\sqrt{1-|z_k|^2}|B_k'(z_k)|} T_k(z) \quad . \tag{5.5}$$

The condition is the convergence of the series:

$$\sum_{k=0}^{\infty} \frac{|y_k-F_{k-1}(z_k)|^2}{(1-|z_k|^2)|B_k'(z_k)|^2} < \infty \quad . \tag{5.6}$$

The condition (5.6) determines a set in $\mathbb{C}^\infty$, called the <u>admissible set of interpolation data</u>, that is, the values that a function in H^2 can assume at the points $\{z_i\}^\infty_{i=0}$. This set will be denoted by <u>R</u>.

The uniqueness of the interpolating function (5.2) or (5.5) depends on the nature of the abscissae.

Walsh [10; pp. 305-306] proves that if the set of interpolation data Y is in R and if the abscissae $\{z_i\}^\infty_{i=0} \in B_1$ are such that $\prod_{i=0}^{\infty} |z_i|$ diverges to zero, then the function F(z) given by (5.2) or (5.5) is unique. If the interpolation data Y is in R and if the product of abscissae $\prod_{i=0}^{\infty} |z_i|$ converges, then (5.2) and (5.5) converge interior to $|z| = 1$ (and in the mean on $|z| = 1$) to the function $F(z) \in H^2$ of minimum norm which satisfies the required conditions of interpolation.

In the first case $\{T_k(z)\}_{k=0}^{\infty}$ is a complete set so that (5.5) is the unique Fourier expansion of the interpolant. On the other hand, in the second instance $\{T_k(z)\}_{k=0}^{\infty}$ is incomplete; (5.5) provides other unique minimum norm interpolant.

For the abscissae $\{z_i\}_{i=0}^{\infty}$, the operator,

$$\Lambda : H^2 \to \mathbb{C}^{\infty}$$
$$\Lambda f = \{f(z_i)\}_{i=0}^{\infty} \tag{5.7}$$

where $\mathbb{C}^{\infty}$ is the class of sequences of complex numbers, will be called the <u>infinite interpolation operator</u>.

<u>The Generalized Inverse Solution</u>

Since the nonunique case is the more general, we will assume that $\prod_{i=0}^{\infty} |z_i|$ diverges to zero. The Takenaka-Walsh solution (5.2) or equivalently (5.5) will be shown to be a generalized inverse solution.

To the orthonormal sequence $\{T_k(z)\}_{k=0}^{\infty}$, we adjoin the functions $\{W_k(z)\}_{k=0}^{\infty}$ so that the two sequences form a complete orthonormal system.

The operator X, that maps admissible sequences of data Y into the minimum norm interpolating functions in H^2,

$$X : R \to H^2$$
$$X = \sum_{k=0}^{\infty} \frac{B(z)}{(z-z_k)B'(z_k)} e_k \tag{5.8}$$

will be called the <u>Takenaka-Walsh solution operator</u>.

We note that X is defined only on $R \subset \mathbb{C}^{\infty}$, however by defining a suitable inner product, R can be shown to be itself a Hilbert space.

<u>Theorem 7</u>. The following are properties of R, the range of Λ:

(i) R is a linear manifold,

(ii) $(Y_1, Y_2)_R = (XY_1, XY_2)_{H^2}$ is an inner product on R (5.9)

(iii) R is a Hilbert space.

<u>Proof</u> (i) For $Y_1 = \{y_k^1\}_{k=0}^{\infty}$ and $Y_2 = \{y_k^2\}_{k=0}^{\infty} \in R$ there corresponds an f_1 and f_2 such that $Y_1 = \{f_1(z_k)\}_{k=0}^{\infty}$ and $Y_2 = \{f_2(z_k)\}_{k=0}^{\infty}$. Thus for the complex scalars, α and β, $\alpha Y_1 + \beta Y_2 = \{\alpha f_1(z_k) + \beta f_2(z_k)\}_{k=0}^{\infty}$ $= \Lambda(\alpha f_1 + \beta f_2) \in R$.

(ii) follows upon direct verification of the four conditions characterizing an inner product.

(iii) let $\{Y_n\}_{n=0}^{\infty}$ be Cauchy in R, then

$||Y_n - Y_m||^2 = (Y_n - Y_m, Y_n - Y_m) = (F_n - F_m, F_n - F_m)_{H^2} = ||F_n - F_m||^2_{H^2} \to 0$

as $n, m \to \infty$ where $F_n = XY_n$ and $F_m = XY_m$ are elements in the subspace M of H^2 spanned by $\{T_k(z)\}_{k=0}^{\infty}$. Since $\{F_n\}$ is Cauchy in a subspace M we have that $\lim_{n\to\infty} F_n = F \in M$. Corresponding to F is the element Y in R, $\Lambda F = Y$, such that $\lim_{n\to\infty} Y_n \to Y \in R$. □

Now we consider the infinite interpolation operator in light of the topology on R.

<u>Theorem 8</u>. The operator Λ has the following properties.

(i) for any $f \in H^2$

$$||f||_R = ||F||_{H^2} \tag{5.10}$$

where F is the minimum norm function in H^2 such that $F(z_i) = f(z_i)$, $i=0,1,\ldots,$

(ii) Λ is a bounded linear operator with respect to R. Λ induces the orthogonal decomposition of H^2.

$$H^2 = N(\Lambda) \oplus N^{\perp}(\Lambda) \tag{5.11}$$

with

$$N^{\perp}(\Lambda) = \bigvee_{k=0}^{\infty} T_k(z) \tag{5.12}$$

$$N(\Lambda) = \bigvee_{k=0}^{\infty} W_k(z) \tag{5.13}$$

<u>Proof</u> From the definition of the inner product (5.9), for any $f \in H^2$ we have that $||\Lambda f||_R^2 = (Y,Y)_R = (XY,XY)_{H^2} = ||F||^2_{H^2}$. Since X maps the

vector of interpolation data into the minimum norm interpolating function,

$$F(z_i) = f(z_i) \qquad i=0,1,\dots,$$
$$||F|| \le ||f|| \,. \tag{5.14}$$

(ii) From (5.10) and (5.14), $||\Lambda f||_R = ||F||_{H^2} \le ||f||_{H^2}$, we have that Λ is a bounded operator and is closed since Λ is defined on all of H^2.

(iii) From

$$N(\Lambda) = \{f_o \in H^2 : (f_o(z), K(z,z_i)) = 0, \quad i=0,1,\dots,\}$$
$$= \{f_o \in H^2 : (f_o(z), T_i(z)) = 0, \quad i=0,1,\dots,\}$$

we have that $\{W_k(z)\}_{k=0}^{\infty}$ spans $N(\Lambda)$ and $\{T_k(z)\}_{k=0}^{\infty}$ spans $N^{\perp}(\Lambda)$ and $H^2 = N(\Lambda) \oplus N^{\perp}(\Lambda)$. □

We are now in a position to prove that the Takenaka-Walsh solution operator (5.8) is a restricted generalized inverse of Λ.

Theorem 9. The Takenaka-Walsh solution operator X (5.8) satisfies the four Moore-Penrose conditions (2.4) when Y is restricted to the subspace $R \subset \mathbb{C}^{\infty}$ and is as a result a restricted generalized inverse of Λ.

Proof For any $f \in H^2$, $\Lambda X \Lambda f = \Lambda XY = \Lambda F = \Lambda f$. For any $Y \in R$, $X\Lambda XY = X\Lambda F = XY$. For $Y_1, Y_2 \in R$, $(\Lambda XY_1, Y_2) = (X\Lambda XY_1, XY_2) = (XY_1, XY_2) = (XY_1, X\Lambda XY_2) = (Y_1, \Lambda XY_2)$. Hence $(\Lambda X)^* = \Lambda X$. For $f_1, f_2 \in H^2$,

$$(X\Lambda f_1, f_2) = (X\Lambda P_{N(\Lambda)} f_1, P_{N(\Lambda)} f_2) + (X\Lambda P_{N^{\perp}(\Lambda)} f_1, P_{N(\Lambda)} f_2)$$
$$+ (X\Lambda P_{N(\Lambda)} f_1, P_{N^{\perp}(\Lambda)} f_2) + (X\Lambda P_{N^{\perp}(\Lambda)} f_1, P_{N^{\perp}(\Lambda)} f_2)$$
$$= 0 + 0 + 0 + (P_{N^{\perp}(\Lambda)} f_1, X\Lambda P_{N^{\perp}(\Lambda)} f_2)$$
$$= (P_{N(\Lambda)} f_1, X\Lambda P_{N(\Lambda)} f_2) + (P_{N^{\perp}(\Lambda)} f_1, X\Lambda P_{N(\Lambda)} f_2)$$
$$+ (P_{N(\Lambda)} f_1, X\Lambda P_{N^{\perp}(\Lambda)} f_2) + (P_{N^{\perp}(\Lambda)} f_1, X\Lambda P_{N^{\perp}(\Lambda)} f_2)$$
$$= (f_1, X\Lambda f_2).$$

Thus $(X\Lambda)^* = X\Lambda$. □

In light of Theorem 9, the Takenaka-Walsh solution operator (5.8) will be denoted by $X = \Lambda_r^\dagger$. The minimum norm interpolating function for the given data Y is then $\Lambda_r^\dagger Y$.

Recalling that $F(z) = \Lambda_r^\dagger Y$ is a linear combination of $\{K(z,z_k)\}_{k=0}^\infty$, there is a relationship between a reproducing kernel and the restricted generalized inverse, $\Lambda_r^\dagger$.

The reproducing kernel for H^2 can be formed as

$$K(z,t) = \sum_{k=0}^{\infty} T_k(z)\overline{T_k(t)} + \sum_{k=0}^{\infty} W_k(z)\overline{W_k(t)} \tag{5.15}$$

since $\{T_k(z)\}_{k=0}^\infty \cup \{W_k(z)\}_{k=0}^\infty$ is a complete set for H^2.

For the subspace $N^\perp(\Lambda)$ we define the kernel using the Takenaka-Walsh sequence $\{T_k(z)\}_{k=0}^\infty$

$$K_{N^\perp(\Lambda)}(z,t) = \sum_{i=0}^{\infty} T_i(z)\overline{T_i(t)} \tag{5.16}$$

Any $f \in H^2$ has the decomposition $f = P_{N^\perp(\Lambda)}f + P_{N(\Lambda)}f$. Then

$$(f(z),K_{N^\perp(\Lambda)}(z,t)) = P_{N^\perp(\Lambda)}f(z) = F(t) \tag{5.17}$$

where $F(t)$ is the Takenaka-Walsh solution such that $F(z_i) = f(z_i)$, $i=0,1,\ldots$.

From the projection properties of Λ and $\Lambda_r^\dagger$

$$\Lambda_r^\dagger\Lambda = P_{N^\perp(\Lambda)} \tag{5.18}$$

so that for f and F in (5.17), $\Lambda_r^\dagger\Lambda f = F$. Thus the following theorem has been proved.

Theorem 10. The reproducing kernel on $N^\perp(\Lambda)$ projects a function f in H^2 onto its minimum norm counterpart in $N^\perp(\Lambda)$, that is

$$(f(z),K_{N^\perp(\Lambda)}(z,t)) = F(t)$$

$$f(z_i) = F(z_i), \quad i=0,1,\ldots,$$

$$||f|| > ||F|| .$$

The identification $\Lambda_r^\dagger\Lambda f = (f(z),K_{N^\perp(\Lambda)}(z,t)) = F(z)$ is also possible.

Another Interpretation

The calculation of the Fourier coefficients of the expansion (5.5) in terms of the Takenaka-Walsh sequence is dependent upon a process of interpolation. We now present the relationship between the interpolation conditions and the interpolation series (5.5). It will be the generalized inverse viewpoint applied to the works of Walsh and Davis [4], [12].

The interpolation problem we consider is the general case of infinite interpolation to any set of bounded linear functional conditions on a Hilbert space possessing a reproducing kernel.

Let H be such a Hilbert space of functions defined on some region B in the complex plane. Given a set of interpolation data $\{y_i\}_{i=0}^{\infty} = Y \in \mathbb{C}^{\infty}$ and a sequence of independent bounded linear functionals $\{L_i\}_{i=0}^{\infty}$, we seek the function in H such that

$$L_i f = y_i, \quad i=0,1,\ldots, \tag{5.19}$$

$$||f||_H \text{ is a minimum} \tag{5.20}$$

From (4.7) we have that $L_i f = (f(z), \overline{L_{i,t}K(t,z)})$, $i=0,1,\ldots,$. The representers of $\{L_i\}_{i=0}^{\infty}$, $\{\overline{L_{i,t}K(t,z)}\}_{i=0}^{\infty}$ is a set of independent functions in H. We orthonormalize this set by the Gram-Schmidt process to obtain the new sequence $\{\phi_k(z)\}_{k=0}^{\infty}$

$$\phi_k(z) = \sum_{i=0}^{k} c_{ki}\overline{L_{i,t}K(t,z)}, \quad k=0,1,\ldots, \quad . \tag{5.21}$$

A new set of linear functionals $\{L_i^*\}_{i=0}^{\infty}$, can also be produced,

$L_i^* f = (f(z),\phi_i(z))$, $i=0,1,\ldots,$. This set is a linear combination of

$\{L_i\}_{i=0}^{\infty}$, since

$$L_i^* f = (f(z), \sum_{j=0}^{i} c_{ij} \overline{L_{j,t}K(t,z)} = \sum_{j=0}^{i} \overline{c}_{ij} L_i f \quad .$$

The functions $h_j(z)$ and functionals ℓ_i are said to be <u>biorthogonal</u> if $\ell_i h_j(z) = \delta_{ij}$ for all i and j. The orthonormal sequence $\{\phi_k(z)\}_{k=0}^{\infty}$ is simultaneously biorthogonal to the set of linear functionals $\{L_i^*\}_{i=0}^{\infty}$ since, $L_i^* \phi_j(z) = (\phi_j(z),\phi_i(z)) = \delta_{ji}$ for all i and j.

With these sequences the existence of a function satisfying (5.19) can be easily determined and expressed.

If the interpolation data Y is specified such that

$$\sum_{i=0}^{\infty} |L_i^* F|^2 < \infty , \tag{5.22}$$

then the solution to (5.19) exists and is given by

$$F(z) = \sum_{i=0}^{\infty} L_i^* F \;\; \phi_i(z) \quad . \tag{5.23}$$

If $\{\phi_k(z)\}_{k=0}^{\infty}$ is a complete set in H, then (5.23) is the unique interpolating function. If $\{\phi_k(z)\}_{k=0}^{\infty}$ is not complete, then (5.23) is the minimum norm interpolating function (Davis [4]).

An operator similar to the infinite interpolation operator Λ may be defined for this problem. Let Λ' be the operator

$$\Lambda' : H \to R' \subset \mathbb{C}^{\infty} \tag{5.24}$$

$$\Lambda' f = \{L_i f\}_{i=0}^{\infty} \tag{5.25}$$

where $R' \subset \mathbb{C}^{\infty}$ is an admissible set of data determined by (5.22).

The solution (5.22) to the interpolation problem (5.19)-(5.20) can be identified as the restricted generalized inverse solution.

The minimum norm interpolating function (5.23) can be written in matrix form:

$$F(z) = [\phi_o, \phi_1, \ldots] \begin{bmatrix} \bar{c}_{00} & & & \\ \bar{c}_{10} & \bar{c}_{11} & & \\ \bar{c}_{20} & \bar{c}_{21} & \bar{c}_{22} & \\ & & & \ddots \end{bmatrix} \begin{bmatrix} y_o \\ y_1 \\ \vdots \end{bmatrix}$$

The operator X

$$X : R' \to H$$

$$X = [\phi_o, \phi_1, \ldots] \begin{bmatrix} \bar{c}_{00} & & & \\ \bar{c}_{10} & \bar{c}_{11} & & \\ \bar{c}_{20} & \bar{c}_{21} & \bar{c}_{22} & \\ & \vdots & & \ddots \end{bmatrix}$$

can easily be shown, as was done in Theorem 9, to be the restricted generalized inverse of Λ' . Thus

$$\Lambda_r^{\dagger} = [\phi_0 \phi_1, \ldots] \begin{bmatrix} \bar{c}_{00} & & & \\ \bar{c}_{10} & \bar{c}_{11} & & \\ \bar{c}_{20} & \bar{c}_{21} & \bar{c}_{22} & \\ & \vdots & & \ddots \end{bmatrix}.$$

REFERENCES

1 Stefan Bergman, The kernal function and conformal mapping, 2nd ed. (Providence, RI: American Mathematical Society, 1970).

2 Rosemary E. Chang, The application of the generalized inverse to interpolation theory. Ph.D. Dissertation, Brown University, 1978.

3 Philip Davis, Interpolation and approximation. (New York: Dover, 1975).

4 Philip Davis, Linear functional equations and interpolation series. Pacific J. Math., 4(1954) 503-532.

5 C. A. Desoer and B. H. Whalen, A note on pseudoinverses. SIAM J., 11(1963)442-447.

6 Einar Hille, Introduction to general theory of reproducing kernels. Rocky Mts. J. Math., 3(1972) 321-368.

7 M. Z. Nashed, Generalized inverses, normal solvability, and iteration for singular operator equations in Nonlinear functional analysis and applications (L.B.Rall, ed.) (New York: Academic Press, 1971)311-359.

8 W. V. Petryshyn, On generalized inverses and on the uniform convergence of $(I-\beta K)^n$ with applications to iterative methods. J. Math. Anal. Appl. 18(1967)417-439.

9 Walter Rudin, Real and complex analysis. (New York: McGraw-Hill, 1966).

10 Satoru Takenaka, On the orthogonal functions and a new formula of interpolation. Japanese J. Math. 2(1925)129-145.

11 J. L. Walsh, Interpolation and approximation by rational functions in the complex domain, 5th ed. (Providence, R.I.: American Mathematical Society, 1969).

12 J. L. Walsh and Philip Davis, Interpolation and orthogonal systems. J. D'Analyse Mathematique, 2(1952-1953)1-28.

Rosemary E. Chang
Sandia National Laboratories
Livermore, California 94550

The author would like to thank Prof. Philip J. Davis under whose direction a dissertation was written on which this paper is based. This work was funded in part by the National Science Foundation through the contracts of NSF MPS 74-07502 and NSF 75-07468.

C W GROETSCH AND B J JACOBS

Iterative methods for generalized inverses based on functional interpolation

ABSTRACT

Functional interpolation is used to develop two iterative methods for computing the generalized inverse of a bounded linear operator. Asymptotic error bounds are derived for both methods for the case of operators with closed range as well as for the case of operators with arbitrary range.

1. INTRODUCTION

Suppose that H_1 and H_2 are Hilbert spaces over the same field of scalars and that $T:H_1 \to H_2$ is a bounded linear operator. Given a vector $b \in H_2$, we say that $u \in H_1$ is a least squares solution of the equation

$$Tx = b \tag{1}$$

if $||Tu-b|| \leq ||Tx-b||$ for all $x \in H_1$. It is not difficult to show that this inequality is equivalent to the conditions

$$Tu = Pb \tag{2}$$

and

$$T^*Tu = T^*b \tag{3}$$

where P is the projection operator of H_2 onto $\overline{R(T)}$ (the closure of the range of T) and T^* is the adjoint of T. If T happens to have closed range, then for each $b \in H_2$, $Pb \in R(T)$ and hence equation (2) has a solution, i.e. for each $b \in H_2$ equation (1) has at least one least squares solution. Since T is linear and continuous it follows from (2) that the set of all least squares solutions is closed and convex and therefore contains a unique vector of smallest norm. The operator, which we will denote by T^+, which associates with each $b \in H_2$ the unique least squares solutions of smallest norm of the equation (1) is called the generalized inverse of T (the term generalized inverse is often used in a wider sense; the operator T^+ which we have defined is commonly called the Moore-Penrose generalized inverse of T). It is not difficult to show that $T^+:H_2 \to H_1$ as defined above is both linear and continuous.

If T does not have closed range, then for certain $b \in H_2$, $Pb \notin R(T)$ and therefore equation (2) has no solution, i.e. equation (1) has no least squares solution for such b. In order to define a generalized inverse as above we therefore must restrict out attention to those $b \in H_2$ which satisfy $Pb \in R(T)$. But this set of vectors b is the set

$$R(T) \oplus R(T)^{\perp} = \{x+y: x \in R(T), \ y \in R(T)^{\perp}\} \quad .$$

Therefore we see that if we are to define a generalized inverse as above, then the largest set which may serve as its domain is the dense subspace $\mathcal{D}(T^+) = R(T) \oplus R(T)^{\perp}$ of H_2. We accordingly define the generalized inverse of T to be the operator $T^+:\mathcal{D}(T^+) \to H_1$ which associates with each $b \in \mathcal{D}(T^+)$ the unique smallest norm least squares solution, T^+b, of the equation (1). Again T^+ turns out to be a linear operator but if $R(T)$ is not closed, T^+ is an unbounded operator (see [6] and [3]).

The dichotomous situation outlined above manifests itself in two ways in the study of approximations to the generalized inverse of a bounded linear operator with nonclosed range. First, approximations to T^+ by bounded linear operators obviously cannot converge in the uniform sense if $R(T)$ is not closed. Also, error bounds for pointwise approximations to T^+ by bounded operators have thus far been obtained only under a special assumption on the vector b.

In [2] and [4] (see also [10] and [3]) spectral theory is used to provide a simple general framework for the study of approximations to the generalized inverse of a bounded linear operator with closed range and in particular it is shown that many well-known approximation techniques for the generalized inverse may be viewed as applications of classical summability theory. In a similar vein we shall investigate in this paper two iterative methods which result when classical functional interpolation methods are lifted via spectral theory to approximate the generalized inverse. The methods are based on Lagrange and Hermite interpolation and asymptotic error bounds are derived both for the case of operators with closed range as well as for the case of operators with arbitrary range.

2. OPERATORS WITH CLOSED RANGE

If $T:H_1 \to H_2$ is a bounded linear operator with closed range, then $\mathcal{H} = R(T^*)$ is a Hilbert space and the operator $\tilde{T}:\mathcal{H} \to \mathcal{H}$ defined by $\tilde{T} = T^*T|\mathcal{H}$ (the restriction of T^*T to $\mathcal{H}$) is an invertible positive definite operator (see e.q. [9]) and it is not difficult to show that

$$T^+ = \tilde{T}^{-1}T^* \quad . \tag{4}$$

It is therefore evident that in approximating $\tilde{T}^{-1}$ one obtains corresponding approximations to T^+. A general result along these lines is the following.

Theorem 1 Suppose that $T:H_1 \to H_2$ is a bounded linear operator with closed range and let $\tilde{T}:\mathcal{H} \to \mathcal{H}$ be the restriction of T^*T to $\mathcal{H} = R(T^*)$. If $\{S_n(x)\}$ is a sequence of continuous real-valued functions on $(0,||T||^2]$ satisfying $\lim_n S_n(x) = x^{-1}$ uniformly on compact subsets of $(0,||T||^2]$, then $T^+ = \lim S_n(T)T^*$ where the convergence is in the operator norm, moreover

$$||T^+ - S_n(\tilde{T})T^*|| \leq \sup_{x \in \sigma(\tilde{T})} |x\, S_n(x) - 1| \quad ||T^+|| \ .$$

Proof First observe that $\sigma(\tilde{T})$ is a compact subset of $(0,||T||^2]$. By equation (4) we see that $T^* = \tilde{T}T^+$ and hence $T^+ - S_n(\tilde{T})T^* = (I - S_n(\tilde{T})\tilde{T})T^+$ where I is the identity operator on $\mathcal{H}$, also $||I - S_n(\tilde{T})\tilde{T}|| = \sup_{x \in \sigma(\tilde{T})} |1-x\, S_n(x)|$ by the spectral mapping theorem and the spectral radius formula for bounded self-adjoint linear operators. Since $||T^+ - S_n(\tilde{T})T^*|| < ||I - S_n(\tilde{T})\tilde{T}||\ ||T^+||$ the result follows. □

Applications of this result to various series, integral, and regularization methods for approximating the generalized inverse are given in [3]. In order to make use of this general error bound it will be helpful to have a lower bound for $\sigma(\tilde{T})$. Note that for $x \in \mathcal{H}$ we have $T^+Tx = \tilde{T}^{-1}T^*Tx = \tilde{T}^{-1}\tilde{T}x = x$. Therefore $||x||^2 \leq ||T^+||^2||Tx|| = ||T^+||^2 \langle\tilde{T}x,x\rangle$. Hence we see that

$$\lambda \geq ||T^+||^{-2} \quad \text{for each} \quad \lambda \in \sigma(\tilde{T}) \ . \tag{5}$$

This provides the needed lower bound.

In [1], J. C. Dunn showed that a certain iterative method for calculating the inverse of a nonsingular normal operator could be regarded as an application of Lagrange interpolation. Here we will use the above theorem and the Lagrange and Hermite approximants to the function $f(x) = x^{-1}$ to generate iterative methods for computing T^+ and corresponding asymptotic error bounds.

If $p_n(x)$ denotes the unique polynomial of degree $\leq n$ which interpolates the function $f(x) = x^{-1}$ at the points $x = 1,2,\ldots,n+1$, then the Gregory-Newton interpolation formula gives the representation

$$p_n(x) = \sum_{k=0}^{n} \binom{x-1}{k} \Delta^k f(1)$$

where $\Delta f(x) = f(x+1) - f(x)$, $\Delta^j f(x) = \Delta(\Delta^{j-1} f)(x)$ and

$\binom{x-1}{k} = \dfrac{(x-1)(x-2)\ldots(x-k)}{k!}$. It is an easy matter to verify that

$\Delta^k f(1) = \dfrac{(-1)^k}{k+1}$ which gives

$$p_n(x) = \sum_{k=0}^{n} \frac{1}{k+1} \prod_{j=0}^{k-1} \left(1 - \frac{x}{1+j}\right) . \tag{6}$$

Here the product from 0 to -1 is, by convention, taken to be 1. A simple inductive argument shows that

$$1 - xp_n(x) = \prod_{j=0}^{n} \left(1 - \frac{x}{1+j} \right) , \quad n = 0,1,2,\ldots \ . \tag{7}$$

Note that for j sufficiently large, say $j \geq J$, we have

$0 < 1 - \dfrac{x}{1+j} \leq \exp(- \dfrac{x}{1+j})$ for all $x \in \sigma(T)$. Therefore

$0 < \prod_{j=J}^{n} (1 - \dfrac{x}{1+j}) \leq \exp(-x \sum_{j=J}^{n} \dfrac{1}{1+j})$, $n \geq J$. Also,

$$\sum_{j=J}^{n} \frac{1}{1+j} \geq \int_{J+1}^{n+2} \frac{dt}{t} = \log(n+2) - \log(J+1)$$

and hence

$$\exp(-x \sum_{j=J}^{n} \frac{1}{1+j}) \leq (J+1)^{x}(n+2)^{-x}$$

and we see that $|1 - xp_n(x)| \leq C(n+2)^{-x}$ for all $x \in \sigma(\tilde{T})$ for a suitable constant C (depending on T). It follows from (6) and (7) and Theorem 1 that if we set $T_n = p_n(\tilde{T})T^*$, then

$$\begin{aligned} ||T_n - T^+|| &\leq \sup_{x \in \sigma(\tilde{T})} |1 - xp_n(x)| \; ||T^+|| \\ &\leq C \, ||T^+|| \; (n+2)^{-||T^+||^{-2}} \quad . \end{aligned}$$

To phrase this result in terms of an iterative sequence we note that if T_n is defined as above, then

$$T_o = T^*, \quad T_{n+1} = T_n - \frac{1}{n+2} [T^*T \; T_n - T^*] \quad . \tag{8}$$

We have therefore proved the following.

Theorem 2 Suppose that $T:H_1 \to H_2$ is a bounded linear operator with closed range and the operators $T_n:H_2 \to H_1$ are defined by (8), then $T_n \to T^+$ in the uniform topology for operators. Moreover, for a certain constant C,

$$||T_n - T^+|| \leq C||T^+||(n+2)^{-||T^+||^{-2}} \quad .$$

We now take the natural next step of investigating the use of Hermite interpolation in obtaining approximate methods for the generalized inverse. We seek the unique polynomial $q_n(x)$ of degree $\leq$ 2n+1 which satisfies

$$q_n(i) = \frac{-1}{i} \quad , \quad i = 1,2,\ldots,n+1$$

and

$$q_n'(i) = \frac{-1}{i^2} \quad , \quad i = 1,2,\ldots,n+1 \quad .$$

We claim that

$$q_n(x) = \sum_{i=0}^{n} (2(i+1) - x) \prod_{j=1}^{i} \left(\frac{j-x}{j+1} \right)^2 , \quad n = 0,1,\ldots \quad , \tag{9}$$

but choose not to subject the reader to the details of the derivation. An easy inductive argument gives

$$1 - xq_n(x) = \prod_{j=0}^{n} \left(1 - \frac{x}{1+j}\right)^2 , \quad n = 0,1,\ldots \tag{10}$$

which leads to the recurrence formula

$$q_{n+1}(x) = q_n(x) + \frac{1}{n+2}\left(2 - \frac{1}{n+2}\right)(1 - xq_n(x)) \tag{11}$$

where $q_0(x) = 2-x$. Hence if we set $T_n = q_n(\tilde{T})T^*$, we have the following iterative method for computing the generalized inverse:

$$\begin{aligned} T_o &= 2T^* - T^*TT^* \\ T_{n+1} &= T_n - \frac{1}{n+2}\left(2I_1 - \frac{1}{n+2} T^*T\right)(T^*TT_n - T^*) \end{aligned} \tag{12}$$

where I_1 is the identity operator on H_1. Following the discussion of the previous method and using (10) we find that for a certain constant K,

$$\sup_{x \in \sigma(\tilde{T})} |1 - xq_n(x)| \leq K(n+2)^{-2}||T^+||^{-2} .$$

Using Theorem 1 we thus obtain

Theorem 3 Suppose that $T:H_1 \to H_2$ is a bounded linear operator with closed range and the operators $T_n:H_2 \to H_1$ are defined by (12), then $T_n \to T^+$ as $n \to \infty$, where the convergence is in the uniform norm. Moreover, for a certain constant K,

$$||T_n - T^+|| \leq K||T^+||(n+2)^{-2}||T^+||^{-2} .$$

3. OPERATORS WITH ARBITRARY RANGE

If $R(T)$ is not closed, then T^+ is an unbounded operator and hence it is impossible to find a sequence of bounded linear operators $\{T_n\}$ such that $T_n \to T^+$ in uniform norm as $n \to \infty$. We must therefore search instead for bounded linear operators $T_n : H_2 \to H_1$ converging pointwise to T^+, i.e. $||T_n b - T^+ b|| \to 0$ as $n \to \infty$ for each $b \in \mathcal{D}(T^+)$.

The authors [4] have shown (see also [10]) that if $\{S_n(x)\}$ is a sequence of continuous real-valued functions on $(0, ||T||^2]$ such that $S_n(x) \to x^{-1}$ for each $x \in (0, ||T||^2]$ and $\{xS_n(x)\}$ is uniformly bounded, then $T^+ b = \lim_n S_n(\tilde{T}) T^* b$ for each $b \in \mathcal{D}(T^\dagger)$, where $T = T^*T\big|_{\overline{R(T^*)}}$. It is easy to show that the sequences of functions defined by (6) and (9) possess the required properties and therefore the two methods of the previous section will converge pointwise to $T^\dagger$ for each $b \in \mathcal{D}(T^+)$. Below we will derive asymptotic error bounds for these sequences under an additional assumption on the vector b.

Observe that if we set $x_n = T_n b$, then we obtain from (8) and (12) the pointwise iterations $x_{n+1} = x_n - A_n r_n$ where $A_n = \frac{1}{n+2} I_1$ for (8) and $A_n = \frac{2}{n+2} I_1 - \frac{1}{(n+2)^2} T^*T$ for (12) and r_n is the residual vector, $r_n = T^*Tx_n - T^*b$. This form is strongly reminiscent of the steepest descent method with the "operator parameter" A_n replacing the optimal step size $\alpha_n = ||r_n||^2/||Tr_n||^2$.

We find it convenient to deal with a slightly more general procedure. Let $T : H_1 \to H_2$ be a bounded linear operator, $b \in \mathcal{D}(T^\dagger)$, $\mathcal{H}_1 = N(T)^\perp$, $\mathcal{H}_2 = N(T^*)^\perp$ and denote by Q the projection of H_2 onto $\mathcal{H}_2 = \overline{R(T)}$. Choose $x_o \in R(T^*T)$ and let

$$r_n = T^*Tx_n - T^*b$$
$$x_{n+1} = x_n - A_n r_n, \quad n = 0,1,2,..., \tag{13}$$

where $A_n = P_n(T^*T|\mathcal{H}_1)$ and P_n is a polynomial with real coefficients.

Define the error vectors e_n by

$$e_n = x_n - T^{\dagger}b \quad . \tag{14}$$

We henceforth make the assumption that $Qb \in R(TT^*)$ and that there exist sequences of positive scalars $\{\alpha_n\}$ and $\{\beta_n\}$ such that for n sufficiently large, say $n \geq n_o$,

$$\beta_n \geq ||T||^2\alpha_n^2 \quad \text{and} \quad \alpha_n I_1 \geq A_n \geq \beta_n I_1 \tag{15}$$

where I_1 is the identity operator on H_1.

Lemma 1 $A_n T^* = T^*\hat{A}_n$, where $\hat{A}_n = P_n(TT^*|H_2)$ and $\alpha_n I_2 \geq \hat{A}_n \geq \beta_n I_2$ for $n \geq n_o$, where I_2 is the identity operator on H_2.

Proof The stated equality may be routinely verified. In order to prove the inequality it is sufficient to show, since A_n and $\hat{A}_n$ are self-adjoint, that $\sigma(A_n) = \sigma(\hat{A}_n)$ (see e.g. [13, p. 320]). By virture of the spectral mapping theorem it is enough to show that $\sigma(T^*T|\ H_1) = \sigma(TT^*|\ H_2)$. If $T_1:H_1 \to H_1$ is defined by $T_1 = T|\ H_1$, then $T_1^* = T^*|\ H_2$ and since $N(T_1) = \{0\}$, there is a self-adjoint operator $B: H_1 \to H_1$ and a unitary operator $U: H_1 \to H_2$ such that $T_1 = UB$ (see [8, p. 283]). It follows that $TT^*|\ H_2 = T_1T_1^* = UB^2U^*$ and $T^*T|\ H_1 = T_1^*T_1 = BU^*UB = B^2$. Therefore the operators $T^*T|\ H_1$ and $TT^*|\ H_2$ are unitarily equivalent and consequently their spectra coincide, which proves the lemma. □

The rest of our argument follows closely those given by Showalter and Ben-Israel [12] and Kammerer and Nashed [5] (see also [7]) for the method of steepest descent.

Lemma 2 $(e_n, A_n r_n) \geq (A_n r_n, A_n r_n)$ for $n \geq n_o$.

Proof First note that $r_n = T^*Te_n$. We then have by Lemma 1 and (15)

$$(e_n, A_n r_n) = (Te_n, \hat{A}_n Te_n) \geq \beta_n ||Te_n||^2$$

$$\geq \alpha_n^2 ||T||^2 ||Te_n||^2 \geq ||T^* \hat{A}_n Te_n||^2$$

$$= (A_n r_n, A_n r_n), \quad \text{for } n \geq n_o . \quad \square$$

<u>Lemma 3</u> $||e_{n+1}||^2 \leq ||e_n||^2 - (e_n, A_n r_n), \quad n \geq n_o$.

<u>Proof</u> Note that $e_{n+1} = e_n - A_n r_n$ by (13) and (14), therefore

$$||e_{n+1}||^2 = (e_n - A_n r_n, e_n - A_n r_n)$$

$$= ||e_n||^2 - 2(e_n, A_n r_n) + (A_n r_n, A_n r_n)$$

$$\leq ||e_n||^2 - (e_n, A_n r_n), \text{ by Lemma 3.} \quad \square$$

By our restriction on the vector b, there exists $\bar{z} \in H_2$ such that $Qb = TT^*\bar{z}$ and without loss of generality we may take $\bar{z} \in \overline{R(T)}$. Since $TT^+b = Qb = TT^*\bar{z}$ and since T is one-to-one on $R(T^*)$, we have $T^+b = T^*\bar{z}$. We observe that $r_n \in R(T^*T)$ and hence $x_n \in R(T^*T)$ for each n. Therefore there are vectors $z_n \in \overline{R(T)}$ such that $x_n = T^*z_n$. Hence $e_n = x_n - T^+b = T^*(z_n - \bar{z})$. It follows that

$$T^*(z_{n+1} - \bar{z}) = e_{n+1} = e_n - A_n r_n$$

$$= T^*(z_n - \bar{z}) - A_n T^*Te_n$$

$$= T^*[I_2 - \hat{A}_n TT^*] (z_n - \bar{z})$$

Since T^* is one-to-one on $R(T)$, we therefore have

$$z_{n+1} - \bar{z} = (I_2 - \hat{A}_n TT^*)(z_n - \bar{z}), \quad n = 0,1,2,\ldots \quad . \tag{16}$$

<u>Lemma 4</u> $||z_{n+1} - \bar{z}|| \leq ||z_n - \bar{z}||$ for $n \geq n_o$.

Proof From (16),

$$||z_{n+1} - \bar{z}||^2 = ||z_n - \bar{z}||^2 - 2(z_n - \bar{z}\ ,\ \hat{A}_n TT^*(z_n - \bar{z}))$$

$$+ (\hat{A}_n TT^*(z_n - \bar{z})\ ,\ \hat{A}_n TT^*(z_n - \bar{z}))\ .$$

Hence it is enough to show that

$$||\hat{A}_n TT^*(z_n - \bar{z})||^2 \leq (z_n - \bar{z},\ \hat{A}_n TT^*(z_n - \bar{z}))\ .$$

This is equivalent to showing that $||TA_n e_n||^2 \leq (e_n, A_n e_n)$.

But $(e_n, A_n e_n) \geq \beta_n ||e_n||^2 \geq ||T||^2 \alpha_n^2 ||e_n||^2 \geq ||TA_n e_n||^2$. □

Lemma 5 $||e_n||^2 \leq ||Te_n||\ ||z_{n_o} - \bar{z}||$ for $n \geq n_o$.

Proof By Lemma 4 we have $||e_n||^2 = (e_n, T^*(z_n - \bar{z})) \leq ||Te_n||\ ||z_n - \bar{z}||$
$\leq ||Te_n||\ ||z_{n_o} - \bar{z}||$. □

Lemma 6 $||e_{n+1}||^2 \leq (1 - \dfrac{\beta_n}{||z_{n_o} - \bar{z}||^2}||e_n||^2)||e_n||^2$ for $n \geq n_o$

Proof

$$\begin{aligned} ||e_{n+1}||^2 &\leq ||e_n||^2 - (e_n, A_n r_n) \\ &= ||e_n||^2 - (Te_n, \hat{A}_n Te_n) \\ &\leq ||e_n||^2 - \beta_n ||Te_n||^2 \\ &\leq ||e_n||^2 - \frac{\beta_n}{||z_{n_o} - \bar{z}||} ||e_n||^4 \end{aligned}$$

by Lemma 5. □

Lemma 7 Suppose $\{\gamma_n\}_{n=n_o}^{\infty}$ and $\{C_n\}_{n=n_o}^{\infty}$ are sequences of numbers satisfying $0 < \gamma_{n_o} C_{n_o} \leq 1$ and $0 < C_{n+1} \leq (1 - \gamma_n C_n) C_n$ for $n \geq n_o$, then

$$C_n \leq C_{n_o} / (1 + C_{n_o} \sum_{k=n_o}^{n-1} \gamma_k)\ ,\qquad n \geq n_o .$$

<u>Proof</u> It is easy to show by induction that $0 < \gamma_n C_n \leq 1$ for $n \geq n_o$. For $k = n_o, n_o+1, \ldots$ we have

$$\frac{1}{C_{k+1}} - \frac{1}{C_k} = \frac{C_k - C_{k+1}}{C_k C_{k+1}} \geq \frac{\gamma_k C_k^2}{C_k C_{k+1}} \geq \gamma_k \quad ,$$

therefore

$$\frac{1}{C_n} - \frac{1}{C_{n_o}} \geq \sum_{k=n_o}^{n-1} \gamma_k$$

which is equivalent to the stated inequality. □

If in the general iterative procedure considered above we set

$\gamma_n = \beta_n ||z_n - z||^{-2}$ and $C_n = ||e_n||^2$, we find that

$$\gamma_{n_o} C_o = \beta_{n_o} \frac{||T^* (z_{n_o} - z)||^2}{||z_{n_o} - \bar{z}||^2} \leq \beta_{n_o} ||T||^2 \quad ,$$

therefore we have:

<u>Theorem 4</u> Suppose $T:H_1 \to H_2$ is a bounded linear operator and $b \in \mathcal{D}(T+)$ satisfies $Qb \in R(TT^*)$. If $\{x_n\}$ is defined by (13) where $\sum\beta_k$ diverges and $\beta_{n_o} \leq ||T||^{-2}$, then $x_n \to T^+b$ as $n \to \infty$.

Moreover

$$||x_n - T^*b|| \leq \frac{||x_{n_o} - T^+b||^2}{1 + \dfrac{||x_{n_o} - T^+b||^2}{||z_{n_o} - \bar{z}||^2} \displaystyle\sum_{k=n_o}^{n-1} \beta_k} \quad , n > n_o \; .$$

In the special case of Lagrange interpolation we have $A_n = \frac{1}{n+2} I_1$ and we may take $\alpha_n = \beta_n = \frac{1}{n+2}$. If we take $n_o \geq ||T||^2 - 2$ we find that $\beta_{n_o} \leq ||T||^{-2}$ and (15) is satisfied for $n \geq n_o$. Therefore the method based on Lagrange interpolation satisfies the hypotheses of Theorem 4. Note also that in this case

$$\sum_{k=n_o}^{n-1} \beta_k \geq \log (n+2) - \log n_o$$

so that $||x_n - T^+b||^2 = 0(\frac{1}{\log n})$ for this method.

For the method based on Hermite interpolation we have

$A_n = \frac{1}{n+2} [2I_1 - \frac{1}{n+2} T^*T]$ and we may take $\alpha_n = \frac{2}{n+2}$ and

$\beta_n = \frac{1}{n+2} [2 - \frac{||T||^2}{n+2}]$. Therefore if we take $n_o \geq \frac{5}{2}||T||^2-2$ we find that $\beta_{n_o} \leq ||T||^{-2}$ and (15) is satisfied for $n > n_o$. In this case it again turns out that

$$||x_n - T^+b||^2 = 0 (\frac{1}{\log n})$$

which shows that, unlike in the closed range case, the asymptotic rate for the second method shows no real improvement over the rate for the method based on Lagrange interpolation.

REFERENCES

1 J. C. Dunn, Inversion of normal operators by polynomial interpolation. Proc. Amer. Math. Soc. 40(1973) 225-228.

2 C. W. Groetsch, Representations of the generalized inverse, J. Math. Anal. Appl. 49(1975) 154-157.

3 C. W. Groetsch, Generalized Inverses of Linear Operators: Representation and Approximation. (Dekker; New York, 1977)

4 C. W. Groetsch and B. Jacobs, A unified convergence theory for generalized inverses of bounded linear operators with arbitrary range. Notices AMS 22(1975) A-681.

5 W. J. Kammerer and M. Z. Nashed, Steepest descent for singular linear operators with nonclosed range. Applicable Anal. 1 (1971) 143-159.

6 W. J. Kammerer and M. Z. Nashed, Iterative methods for best approximate solutions of linear integral equations of the first and second kinds. J. Math. Anal. Appl. 40(1972) 547-573.

7 S. F. McCormick and G. H. Rodrigue, A unified approach to gradient methods for linear operator equations. J. Math. Anal. Appl. 49(1975) 275-285.

8 M. A. Naimark, Normed rings. (Noordhoff:Groningen, 1964) (Translated from the first Russian edition by L. F. Boron).

9 M. Z. Nashed, Steepest descent for singular linear operator equations. SIAM J. Numer. Anal. 7(1970) 358-362.

10 M. Z. Nashed, Perturbations and approximations for generalized inverses and linear operator equations in Generalized inverses and applications (ed. M. Z. Nashed) (New York: Academic Press, 1976).

11 D. W. Showalter, Representation and computation of the pseudoinverse. Proc. Amer. Math. Soc. 18(1967) 584-586.

12 D. W. Showalter, and A. Ben-Israel, Representation and computation of the generalized inverse of a bounded linear operator between Hilbert spaces, Atti. Accad. Naz. Lincei Rend. Cl. Sci. Fis. Mat. Natur., Ser VIII, 48(1970) 120-130.

13 K. Yosida, Functional analysis. Second edition. (Berlin: Springer-Verlag, 1968).

C. W. Groetsch
Department of Mathematical Sciences
University of Cincinnati
Cincinnati, Ohio 45221

B. J. Jacobs
Management Decisions Development Corporation
Cincinnati, Ohio 45240

R H BOULDIN

Generalized inverses and factorizations

INTRODUCTION

Let $T: X \to Y$ be a bounded linear transformation between Hilbert spaces and let y be a fixed vector of Y. Many problems take the form of finding x such that $Tx = y$ is true. If y does not belong to the range of T, then one seeks a best approximate solution u, i.e. $||Tu-y|| = \inf\{||Tx-y|| : x \in X\}$. One usually chooses the best approximate solution with least norm. It is useful to note that u "solves" $Tx = y$ if and only if u solves the associated "normal equation" $T^*Tx = T^*y$. If u is the best approximate solution of $Tx = y$ with least norm and T^+ denotes the Moore-Penrose inverse of T, then $u = T^+y$ (see [20, Theorem 2.1]). Thus, u solves the normal equation if and only if $u = (T^*T)^+T^*y$. The above useful observation is equivalent to the operator equation $T^+ = (T^*T)^+T^*$; this will be called a normal factorization of T^+.

More generally, if $T^\#$ is some generalized inverse of T, then $T^\# = (T^*T)^\#T^*$ and $T^\# = T^*(TT^*)^\#$ will be called the normal factorizations of $T^\#$. The normal factorizations of $T^\#$ will be obtained as consequences of the reverse order law - i.e. $(AB)^\# = B^\#A^\#$. Necessary and sufficient conditions for the reverse order law will be determined relative to the (1,2)-inverse, the Moore-Penrose inverse, and the Drazin inverse.

To define the generalized inverses one usually lists the following properties (see [4]):

$$TXT = T, \quad (1^k)\ T^kXT = T^k \quad , \tag{1}$$

$$XTX = X, \tag{2}$$

$$(TX)^* = TX, \tag{3}$$

$$(XT)^* = XT, \tag{4}$$

$$TX = XT. \tag{5}$$

A (1,2)-inverse of **a** bounded linear transformation T between Banach spaces is a bounded linear transformation satisfying (1) and (2) above; for such a transformation T the Drazin inverse satisfies (5), (2) and (1^k) for some nonnegative integer k; the Moore-Penrose inverse of a bounded linear transformation T between Hilbert spaces is a bounded linear transformation satisfying (1), (2), (3) and (4) above.

Since most factorizations of a generalized inverse arise from factorizations of the transformation which is to be inverted, the approach taken here is to study the reverse order law. In addition to the normal factorizations, polar factorizations, and "full-rank" factorizations, there should be other factorizations of specific transformations that are natural to specific problems. Thus, the reverse order law should be a computational tool as well as a theorem in the basic theory of generalized inverses.

2. THE (1,2)-INVERSE OF THE PRODUCT OF BANACH SPACE TRANSFORMATIONS

The question of existence of (1,2)-inverses for the product AB is settled by well known results. For T a bounded linear transformation (or a densely defined closed linear transformation) between Hilbert spaces, there is a densely defined closed linear (1,2)-inverse, denoted $T^{(1,2)}$; there is a bounded $T^{(1,2)}$ if and only if the range of T is closed (see [21, Theorem 5.6], [3]). In contrast, if $T:X \to Y$ is a bounded linear transformation (or a densely defined closed linear transformation) between Banach spaces, then the existence of $T^{(1,2)}$ requires that the kernel of T, denoted ker T, and the closure of the range of T, denoted $(TX)^-$, are complemented in X and Y, respectively; TX must be closed for the existence of a bounded $T^{(1,2)}$ (see [9], [21]). Having a bounded (1,2)-inverse for T is equivalent to the possibility of solving $Tx = y$ for all y, rather than being restricted to y in a vector space which is dense in Y. Consequently, T is said to be _normally solvable_ provided it has closed range (see [13], [16]).

Following [1], one says that T is _relatively regular_ provided T has a bounded (1,2)-inverse, where T is either bounded or closed with dense domain of definition. The two natural conclusions to seek for the product of relatively regular transformations A and B are that AB is relatively regular and that $(AB)^{(1,2)} = B^{(1,2)}A^{(1,2)}$. Since the first conclusion requires that AB be normally solvable, that is considered next.

The angular distance between two nontrivial subspaces of a Banach space, say M and N, is $\gamma(M,N)$ defined to be

$$\inf\{||x-y|| : x \in M,\ y \in N,\ ||x|| = 1 = ||y||\}$$

and if either M or N is trivial, then $\gamma(M,N)$ is defined to be one. This idea, which seems to originate in [19], is essential in characterizing when AB is normally solvable (see [7]). Let $L(X,Y)$ and $C(X,Y)$ denote the bounded linear transformations and the closed linear transformations, respectively, between the Banach spaces X and Y.

<u>Theorem 2.1</u> Let $A \in L(Y,Z)$, $B \in C(X,Y)$ be normally solvable and let Y_0 denote $\ker A \cap BX$. The product AB is normally solvable if and only if $\gamma(\ker A/Y_0, BX/Y_0)$ is positive.

A useful variant of the preceding theorem is the following corollary.

<u>Corollary 2.2</u> Let $A \in L(Y,Z)$, $B \in C(X,Y)$ be normally solvable and assume $Y_0 = \ker A \cap BX$ is complemented in ker A and BX, say

$$\ker A = Y_1 \oplus Y_0 , \quad BX = Y_2 \oplus Y_0 .$$

Then the following three conditions are equivalent.

The product AB is normally solvable. (1)

$\gamma(Y_1, BX)$ is positive. (2)

$\gamma(\ker A, Y_2)$ is positive. (3)

A characterization of when the product of relatively regular transformations is relatively regular follows from these results (see [7]).

<u>Corollary 2.3</u> Let $A \in L(Y,Z)$, $B \in L(X,Y)$ be relatively regular. The product AB is relatively regular if and only if the following hold:

either $\gamma(Y_1,BX) > 0$ where $\ker A = Y_1 \oplus (\ker A \cap BX)$
or $\gamma(\ker A,Y_2) > 0$ where $BX = Y_2 \oplus (\ker A \cap BX)$, (1)

both ker AB and ABX are complemented in Z. (2).

An alternative characterization was given by J. J. Koliha in [17].

<u>Theorem 2.4</u> Let $A \in L(Y,Z)$, $B \in L(X,Y)$ be relatively regular, let P be a projection onto BX, and let Q be a projection with ker Q equal to ker A. Then AB is relatively regular if and only if QP is relatively regular.

It remains to characterize when $B^{(1,2)}A^{(1,2)}$ is a (1,2)-inverse for AB. In [9] and [17], sufficient conditions like the commutativity of $BB^{(1,2)}$ and $A^{(1,2)}A$ were given; recall that $BB^{(1,2)}$ and $A^{(1,2)}A$ are projections (see [9, Theorem 1]).

Theorem 2.5 Let $A \in L(Y,Z)$, $B \in L(X,Y)$ be relatively regular and denote $BB^{(1,2)}$ and $A^{(1,2)}A$ by P and Q, respectively. The following conditions are equivalent:

$$B^{(1,2)}A^{(1,2)} \text{ is a (1,2)-inverse for } AB, \tag{1}$$

$$A(PQ-QP)B = 0 \text{ and } B^{(1,2)}IPQ-QP)A^{(1,2)} = 0 , \tag{2}$$

$$QP \text{ and } PQ \text{ are projections.} \tag{3}$$

Now the normal factorizations for the (1,2)-inverse of a Hilbert space transformation are given.

Corollary 2.6 If A is a relatively regular bounded linear transformation between Hilbert spaces then the following identities hold:

$$(A^*A)^{(1,2)}A^* = A^{(1,2)}, \tag{1}$$

$$A^*(AA^*)^{(1,2)} = A^{(1,2)}. \tag{2}$$

Proof If T^+ denotes the Moore-Penrose inverse of T, then $A^*(A^*)^+$ is the orthogonal projection onto A^*Y. So ker $A^*(A^*)^+$ is the orthogonal complement of A^*Y, or simply ker A. Thus $A = AA^*(A^*)^+$ and since condition (2) of Theorem 2.5 is readily verified, $((A^*)^+)^{(1,2)}(AA^*)^{(1,2)}$ is a (1,2)-inverse for A. Since A^* is a (1,2)-inverse for $(A^*)^+$, this proves (2) above. Equation (1) follows by applying (2) to A^* and then taking adjoints. □

The analog of the full rank factorization for matrices is now established.

Corollary 2.7 Let A and B be bounded linear transformations between Hilbert spaces. If T = AB where A is one-to-one with closed range and B is onto then the following identities hold:

$$T^{(1,2)} = B^{(1,2)}A^{(1,2)} \tag{1}$$

$$T^{(1,2)} = B^*(BB^*)^{-1}(A^*A)^{-1}A^*. \tag{2}$$

Proof Identity (1) is trivial since $BB^{(1,2)}$ and $A^{(1,2)}A$ are both equal to the identity operator. Identity (2) follows from (1) by using (1) and (2) of Corollary 2.6. □

The proof of (1) above establishes the "full rank" factorization for Banach space transformations. Note that if B is identical with T except that its range space is TX and if A is the identity from TX into Y, then T = AB and the corollary applies.

Other questions that might be considered relate $(AB)^{(1,2)}$ to $A^{(1,2)}$ and $B^{(1,2)}$ without any assumption that A and B have closed range. There might be value in posing each of the preceding questions for generalized inverses in topological vector spaces (see [21]).

3. MOORE-PENROSE INVERSE OF THE PRODUCT OF HILBERT SPACE TRANSFORMATIONS

The existence of the Moore-Penrose inverse has been considered extensively (see [2], [11],[15],[20],[21]). For T a bounded linear transformation (or a densely defined closed linear transformation) between Hilbert spaces there is a unique densely defined closed linear transformation T^+ which is the Moore-Penrose inverse of T; T^+ is bounded if and only if the range of T is closed. It is appropriate to consider Hilbert space conditions for the product of relatively regular transformations to be relatively regular. Note that a Hilbert space transformation is relatively regular if and only if it is normally solvable. Conditions that are necessary and sufficient result from relating angle between subspaces to angular distance between subspaces.

The angle between subspaces M and N, denoted $\alpha(M,N)$, ranges from zero to $\pi/2$ and its cosine is

$$\sup\{|(f,g)|:f \in M, g \in N, \ ||f|| = 1 = ||g||\}.$$

If either M or N is trivial then the angle is $\pi/2$.

<u>Lemma 3.1</u> The angle between subspaces M and N equals zero if and only if the angular distance between M and N, $\gamma(M,N)$, equals zero.

<u>Proof</u> Let $\{f_n\}$ and $\{g_n\}$ be sequences of unit vectors from M and N, respectively, such that $\lim|(f_n,g_n)| = 1$. Choose $\{\alpha_n\}$ such that $|\alpha_n| = 1$ and $(\alpha_n f_n, g_n) = |(f_n,g_n)|$ for $n = 1,2,\ldots$. A simple computation shows $\lim ||\alpha_n f_n - g_n|| = 0$. Thus, $\alpha(M,N) = 0$ implies $\gamma(M,N) = 0$.

If $\{u_n\}$ and $\{v_n\}$ are sequences of unit vectors from M and N, respectively, such that $\lim||u_n - v_n|| = 0$ then it routinely follows that $\lim \mathrm{Re}(u_n,v_n) = 1$. Thus $\gamma(M,N) = 0$ implies $\alpha(M,N) = 0$. □

From Lemma 3.1 and Theorem 2.1 follows a characterization of when the product of transformations with closed range has closed range. Of course, every subspace in a Hilbert space is complemented.

Theorem 3.2 Let $A \in L(Y,Z)$, $B \in C(X,Y)$ be relatively regular; let Y_1 and Y_2 be subspaces of Y such that

$$\ker A = Y_1 \oplus Y_0 , \quad BX = Y_2 \oplus Y_0$$

where $Y_0 = \ker A \cap BX$. Then the following three conditions are equivalent:

the product AB is relatively regular, (1)

$\alpha(Y_1, BX)$ is positive, (2)

$\alpha(\ker A, Y_2)$ is positive. (3)

The next theorem together with the preceding result characterizes when the reverse order law holds.

Theorem 3.3 Let $A \in L(Y,Z)$, $B \in L(X,Y)$ be relatively regular. In order for

$$(AB)^+ = B^+A^+$$

to hold, it is necessary and sufficient that the following conditions hold:

AB is relatively regular, (1)

A^+A commutes with BB^*, and (2)

BB^+ commutes with A^*A. (3).

Proof The proof of Theorem 3.1 of [6] extends from operators to transformations without modification. Thus it suffices to show (2) and (3) are equivalent to the following:

A^*Z is invariant under BB^*, and (4)

$A^*Z \cap \ker B^*$ is invariant under A^*A. (5)

Recall that M is invariant under a self adjoint operator if and only if M reduces it. Also, M reduces T if and only if the orthogonal projection onto M, denoted P_M, commutes with T. Hence, (4) and (5) are equivalent to

P_{A^*Z} commutes with BB^*, and (4')

$P_{A^*Z \cap \ker B^*}$ commutes with A^*A. (5')

It will be shown that (4') and (5') imply (2) and (3). It is seen routinely that $BB^*Y = BX$ and $\ker BB^* = \ker B^*$. Since P_{A^*Z} and $I - P_{A^*Z} = P_{\ker A}$

commute with BB^*, it follows that

$$P_{A^*Z}\, BX \subset BX \cap A^*Z \quad \text{and} \quad P_{\ker A} BX \subset BX \cap \ker A.$$

Note that

$$BX = BX \cap A^*Z \oplus BX \cap \ker A \tag{6}$$

follows from the above containments. Analogously using (5') one obtains

$$P_{A^*Z \cap \ker B^*}\, A^*Z \subset A^*Z \cap \ker B^*, \text{ and}$$

$$P_{\operatorname{span}\{\ker A, BX\}} A^*Z \subset A^*Z \cap \operatorname{span}\{\ker A, BX\}.$$

These containments give a decomposition of A^*Z analogous to (6). Decomposition (6) itself shows that

$$\operatorname{span}\{\ker A, BX\} = \operatorname{span}\{\ker A, BX \cap A^*Z\}$$

and it routinely follows that

$$A^*Z \cap \operatorname{span}\{\ker A, BX\} = A^*Z \cap BX.$$

Thus, it is proved that

$$A^*Z = A^*Z \cap \ker B^* \oplus A^*Z \cap BX. \tag{7}$$

Since A^*Z reduces A^*A, (5') and (7) imply that $A^*Z \cap BX$ reduces A^*A. Since A^*A is zero on $\ker A \cap BX$, (6) now shows that BX is invariant under A^*A. This is equivalent to the commutativity of P_{BX} and A^*A. Since BB^+ equals P_{BX}, one sees that (3) is proved. Statement (2) is immediate.

Now it is shown that (2) and (3) imply (4') and (5'). Clearly (2) implies (4') and (3) implies that BX reduces A^*A, since $BB^+ = P_{BX}$. Because A^*Z obviously reduces A^*A and because $BX = (\ker B^*)^\perp$, one sees that (5') follows. □

As a consequence of the preceding theorem, one easily obtains the normal factorizations for the Moore-Penrose inverse of a transformation.

Theorem 3.4 If T is a relatively regular bounded linear transformation between Hilbert spaces then the following formulas hold:

$$T^+ = (T^*T)^+T^*, \text{ and} \tag{1}$$

$$T^+ = T^*(TT^*)^+. \tag{2}$$

Proof The first formula is derived in [6] from the reverse order law and the second formula is derived analogously starting with the formula $T = TT^*(T^*)^+$. □

For an alternative derivation of the preceding theorem and its subsequent corollary see [15].

Corollary 3.5 Let A and B be bounded linear transformations between Hilbert spaces. If $T = AB$ where A is one-to-one with closed range and B is onto then the following identities hold:

$$T^+ = B^+A^+ \tag{1}$$

$$T^+ = B^*(BB^*)^{-1}(A^*A)^{-1}A^*. \tag{2}$$

Proof Formula (1) follows from Theorem 3.3 by noting that BB^+ and A^+A are both equal to the identity operator and $ABX = AY$ is closed. Formula (2) follows from (1) by using both parts of Theorem 3.4. □

Note that formula (2) of Corollary 2.7 follows from formula (2) above since the Moore-Penrose inverse is certainly a (1,2)-inverse.

4. DRAZIN INVERSE OF THE PRODUCT OF BANACH SPACE OPERATORS

Although the Drazin inverse for a matrix always exists [12], for an operator T on a Banach space X the existence of a Drazin inverse T^D requires that T have finite ascent and finite descent [18, Theorem 6]. Another condition equivalent to the existence of T^D is that the origin is a pole for the resolvent operator of T [9, pp. 51-53]. The existence of A^D and B^D does not imply the existence of $(AB)^D$, and when $(AB)^D$ exists it does not necessarily equal B^DA^D. Before the fundamental theorem of this section is stated, an informative example of the preceding assertion is detailed.

Example 4.1 For $j = 1,2,\ldots$ let H_j be the Hilbert space with orthonormal basis $\{\ldots,e_{-1}(j),e_0(j),e_1(j),\ldots\}$ and let H be the formal direct sum $\sum_j H_j$. For $j = 1,2,\ldots$ define the operator B_j on H_j by letting it be the following matrix on $\text{span}\{e_0(j),e_1(j)\}$ and letting it be zero on the other basis vectors:

$$\begin{bmatrix} 0 & 0 \\ 1/j & 0 \end{bmatrix}.$$

For $j = 1,2,\ldots$ define U_j to be the adjoint of the bilateral shift on H_j relative to the given basis - i.e. $U_j e_i(j) = e_{i-1}(j)$ for all i. Define A to be the formal direct sum $\sum_j U_j$ and define B to be the formal direct sum $\sum_j B_j$. It is verified routinely that AB is zero on all basis vector except $\{e_0(1), e_0(2), \ldots\}$ and $AB\, e_0(j) = (1/j) e_0(j)$. Hence, AB is the orthogonal direct sum of a zero operator and the diagonal operator diag $(1,1/2,1/3,\ldots)$. Since zero is not an isolated point of the spectrum of AB, it is certainly not a pole of the resolvent operator. Thus, $(AB)^D$ does not exist.

The preceding example exposes many of the obstacles in determining conditions necessary and sufficient for the existence of $(AB)^D$, given that A^D and B^D exist. In the example B is nilpotent, in fact, $B^2 = 0$; thus one knows that $B^D = 0$. Since A is unitary, $A^D = A^{-1} = A^*$. This shows that AB can be pathological although A and B are very nice.

With proper hypotheses some strong conclusions are possible. The following theorem is the fundamental result of this section. First an essential lemma is established. Although the lemma is probably well known, a satisfactory reference could not be found.

Lemma 4.2 Let Q_1 and Q_2 be two commuting projections defined on the Banach space X. Then Q_1Q_2 is a projection; furthermore, $Q_1Q_2X = Q_1X \cap Q_2X$ and $(I-Q_1Q_2)X = \ker Q_1Q_2 = \text{span}\{\ker Q_1, \ker Q_2\}$. Consequently the following decomposition holds

$$X = \text{span}\{\ker Q_1, \ker Q_2\} \oplus Q_1X \cap Q_2X.$$

Proof Trivially Q_1Q_2 is idempotent and $Q_1X \cap Q_2X$ contains $Q_1Q_2X = Q_2Q_1X$. Since $Q_1|Q_1X$ and $Q_2|Q_2X$ are the identity maps, Q_1Q_2X contains $Q_1X \cap Q_2X$ and equality is proved. Trivially $\ker Q_1Q_2$ contains $\text{span}\{\ker Q_1, \ker Q_2\}$. Note that

$$X = (I-Q_2)X \oplus Q_1Q_2X \oplus (I-Q_1)Q_2X$$

since $Q_2X = Q_1Q_2X \oplus (I-Q_1)Q_2X$. Because $\ker Q_2$ contains $(I-Q_2)X$ and $\ker Q_1$ contains $(I-Q_1)Q_2X$, the following decomposition must hold

$$X = \text{span}\{\ker Q_1, \ker Q_2\} \oplus Q_1X \cap Q_2X.$$

It follows that $\ker Q_1Q_2 = \text{span}\{\ker Q_1, \ker Q_2\}$. □

Theorem 4.3 Let A and B be (bounded linear) operators on X (a Banach space) with Drazin inverses A^D and B^D, respectively. If B^DB commutes with A, A^DA commutes with B and for some nonnegative integer j

$$\ker(AB)^j \supset \ker A^D \cup \ker B^D \qquad (*)$$

then the following are true:

$$(AB)^D = B^DA^D, \qquad (1)$$

$$\ker(AB)^D = \text{span}\{\ker A^D, \ker B^D\}, \qquad (2)$$

$$(AB)^DX = A^DX \cap B^DX, \text{ and} \qquad (3)$$

the least j for which (*) holds is the ascent and the descent of AB. (4)

Proof Let Q_1 and Q_2 be the projections A^DA and B^DB, respectively. Since the Drazin inverse of T belongs to the second commutant of T [12, Theorem 1], the above hypothesis implies that $A^DQ_2 = Q_2A^D$ and $B^DQ_1 = Q_1B^D$. The following identity results

$$Q_1Q_2 = A^DAQ_2 = Q_2A^DA = Q_2Q_1 .$$

The preceding lemma now applies to Q_1Q_2.

By [18, Theorem 6] the ascent of A equals the descent of A which is a nonnegative integer, say m; furthermore, $\ker A^D$ equals $\ker A^m$ and A^DX equals A^mX. Analogously, $\ker B^D$ equals $\ker B^k$ and B^DX equals B^kX where k is both the ascent and the descent of B. It is elementary that A restricted to A^mX, denoted $A|A^mX$, is bijective and so $AA^DX = A^DX$. Clearly $\ker AA^D \supset \ker A^D$ and since $A|A^DX$ is one-to-one, equality actually holds in the containment. This shows that $\ker Q_1 = \ker A^D = \ker A^m$ and $Q_1X = A^DX = A^mX$. Analogously, $\ker Q_2 = \ker B^D = \ker B^k$ and $Q_2X = B^DX = B^kX$. This establishes the decomposition

$$X = \text{span}\{\ker A^m, \ker B^k\} \oplus A^mX \cap B^kX. \qquad (i)$$

Let p be any positive integer not less than j and note that (*) implies

$$\text{span}\{\ker A^m, \ker B^k\} \subset \ker (AB)^p. \qquad (ii)$$

Because $A|A^mX$ and $B|B^kX$ are bijective and because Q_1 and Q_2 commute with A and B, it follows that $(AB)^p|A^mX \cap B^kX$ is bijective. Thus, one has

$$\ker(AB)^p \cap A^mX \cap B^kX = \{0\} \;;$$

this with (i), (ii) above implies

$$\ker(AB)^p = \operatorname{span}\{\ker A^m, \ker B^k\} \;. \tag{iii}$$

Also, one may conclude that

$$A^mX \cap B^kX \subset (AB)^pX \;. \tag{iv}$$

Note that (iii) shows the ascent of AB does not exceed j.

It is now possible to verify that B^DA^D is the Drazin inverse for AB. Observe that $ABB^DA^D = AQ_2A^D = Q_2AA^D = Q_2Q_1$. Since $(I-Q_1Q_2)X = \ker(AB)^j$ by (iii) and the observations in the first two paragraphs of this proof, one has

$$0 = (AB)^j(Q_2Q_1-I) = (AB)^j(ABB^DA^D-I)$$

$$\text{or} \quad (AB)^{j+1}B^DA^D = (AB)^j \;.$$

Note that

$$ABB^DA^D = Q_2Q_1 = B^DBQ_1 = B^DQ_1B = B^DA^DAB$$

and

$$B^DA^DABB^DA^D = B^DQ_1Q_2A^D = B^DQ_2Q_1A^D$$

$$= B^DBB^DA^DAA^D = B^DA^D \;.$$

From the uniqueness of the Drazin inverse and its definition in [18,Theorem 6], it follows that

$$X = \ker(AB)^D \oplus (AB)^DX = \operatorname{span}\{\ker A^m, \ker B^k\} \oplus (AB)^pX.$$

In view of (i) and (iv) one concludes that

$$(AB)^pX = A^mX \cap B^kX.$$

Since AB has a Drazin inverse, its descent equals its ascent and the theorem is proved. □

It should be remarked that the converse of the preceding theorem does not hold. Let A be a nonzero nilpotent operator on the Hilbert space H and let B be the orthogonal projection onto the orthogonal complement of ker A - i.e. $(\ker A)^{\perp}$. Note that $AB = A$, $A^D = 0$, $(AB)^D = 0$ and $B^D = B$. Although $(AB)^D = B^D A^D$ in this instance, $B^D B = B^2 = B$ does not commute with A. Otherwise $BH = (\ker A)^{\perp}$ would be invariant under A and so $AH = A(\ker A)^{\perp}$ would be contained in $(\ker A)^{\perp}$. This implies that $\ker A^2$ equals $\ker A$; this contradicts that A is nonzero. For a concrete realization of the preceding situation define A and B by

$$A = \begin{bmatrix} 0 & 0 & 0 \\ 1 & 0 & 0 \\ 0 & 1 & 0 \end{bmatrix}, \quad B = \begin{bmatrix} 1 & 0 & 0 \\ 0 & 1 & 0 \\ 0 & 0 & 0 \end{bmatrix} .$$

Provided A is invertible the converse to Theorem 4.3 does hold. In fact, the next theorem establishes a very strong partial converse to Theorem 4.3.

Theorem 4.4 Let B be an operator with a Drazin inverse and let A be an invertible operator. The equation

$$(AB)^D = B^D A^{-1} \qquad (*)$$

is equivalent to

$$B^D B \text{ commutes with } A, \text{ and} \qquad (1)$$

$$\text{for some nonnegative integer } j,\ \ker(AB)^j \text{ contains } \ker B^D. \qquad (2)$$

Proof Since $A^D A = A^{-1} A = I$ commutes with B and $\ker A^D = \{0\}$, Theorem 4.3 shows that (*) follows from (1) and (2).

By the second paragraph of the proof of Theorem 4.3, it is seen that $B^D B$ is the projection onto $B^D X$ along $\ker B^D$. Assume (*) holds and note that

$$AB^D BA^{-1} = (AB)(B^D A^{-1}) = B^D A^{-1} AB = B^D B.$$

This proves (1). Since AB has a Drazin inverse, the ascent and the descent equal the same integer j. Moreover, $(AB)^{j+1} B^D A^{-1} = (AB)^j$ implies that

$$0 = (AB)^j (ABB^D A^{-1} - I) = (AB)^j (B^D B - I).$$

Thus, $\ker(AB)^j$ contains $(I - B^D B)X = \ker B^D B = \ker B^D$ and the theorem is proved. □

It should be observed that condition (2) of Theorem 4.4 can be satisfied in a variety of ways. An equivalent condition is that

$$(AB)^j (I - B^D B) = 0.$$

If ascent of B is one then this condition is necessarily true since

$$(AB)^j - (AB)^{j-1}ABB^DB = (AB)^j - (AB)^{j-1}AB = 0.$$

Condition (2) is satisfied if $\ker B^i$ is invariant under A for $i=1,2,\ldots$. This follows from the fact that $B \ker B^i \subset \ker B^{i-1}$. If B is nilpotent, then condition (2) is equivalent to the statement that AB is nilpotent. Analogous remarks can be made about condition (*) of Theorem 4.3.

The next corollary puts Theorem 4.4 in a form suitable for the application which follows.

Corollary 4.5 Let A and B be operators on X with Drazin inverses and assume $cA + B$ is invertible for some number c. Furthermore, assume the following conditions hold:

A^DA commutes with B, (i)

B^DB commutes with A, (ii)

$\ker([cA+B]^{-1}A)^j$ contains $\ker A^D$, and (iii)

$\ker([cA+B]^{-1}B)^j$ contains $\ker B^D$ for some nonnegative integer j. (iv)

Then the following formulas hold:

$$([cA+B]^{-1}A)^D = A^D[cA+B], \text{ and} \qquad (1)$$

$$([cA+B]^{-1}B)^D = B^D[cA+B]. \qquad (2)$$

Proof As the second paragraph of the proof of Theorem 4.3 showed, A^DA is the projection onto A^DX along $\ker A^D$ and B^DB projects onto B^DX along $\ker B^D$. Since A and B both commute with A^DA it follows that $cA+B$ and $(cA+B)^{-1}$ also commute with A^DA; thus $(cA+B)^{-1}B$ commutes with A^DA. Analogously one shows that $(cA+B)^{-1}A$ commutes with B^DB. These observations and (iii), (iv) above permit two applications of Theorem 4.4 which proves this corollary. □

Corollary 4.6 Assume that the hypothesis of the preceding corollary hold for square matrices A and B. The initial value problem

$$Ax' + Bx = f \quad , \quad x(0) = x_0$$

has a solution for exactly x_0 of the form

$$x_0 = A^DAq + (I-A^DA) \sum_{n=0}^{k-1} (-1)^n (B^DA)^n B^D f^{(n)}(0)$$

for some vector q. A particular solution of $Ax' + Bx = f$ is

$$x = A^D e^{-A^DBt} \int_a^t e^{A^DBs} f(s)ds + (I-A^DA) \sum_{n=0}^{k-1} (-1)^n (B^DA)^n B^D f^{(n)}$$

where a is arbitrary. The general solution of $Ax' + Bx = f$ is the sum of the above particular solution and $e^{-A^DBt}A^DAq$. The solution satisfying $x(0) = x_0$ is found by setting $q = x_0$ and $a = 0$ in the above formulas.

Proof The above statements are a simplification of [8 Theorem 7] which follows from Corollary 4.5. □

Because this last result and its antecedent, Corollary 4.5, seem to have very complicated hypotheses, it might be useful to offer an example where the hypotheses are satisfied. In the next example it is routinely verified that the hypotheses of Corollary 4.5 are trivially satisfied.

Example 4.7 Let B be any invertible matrix in lower triangular form and let A be any nilpotent matrix in lower triangular form. Since the lower triangular form of any matrix enumerates the eigenvalues down the diagonal and since each eigenvalue of a nilpotent matrix is zero, the eigenvalues of cA+B coincide with the eigenvalues of B. It follows that cA+B is lower triangular with only nonzero diagonal entries and, hence, cA+B is invertible. It is easily seen that $(cA+B)^{-1}A$ is nilpotent and, consequently, the zero operator is the Drazin inverse for both it and A. Since B is invertible, $B^D = B^{-1}$ and $B^DB = I$. Thus, each condition in the hypothesis of Corollary 4.5 (and thus also Corollary 4.6) is trivially satisfied.

Although the preceding results give a rather complete characterization of when $(AB)^D$ equals B^DA^D, they do not characterize when $(AB)^D$ exists. In fact, such conditions as the commutativity of B^DB and A cannot be necessary for the existence of $(AB)^D$. Choose X to be finite dimensional, choose B so that B^DB is a nontrivial projection and choose A so that it does not commute with B^DB. Nevertheless, $(AB)^D$ exists. Characterizing the existence of $(AB)^D$ is an interesting unsolved problem.

Finally, this section is concluded by mentioning a factorization of the

Drazin inverse given in [10] for matrices. Unlike earlier formulas this factorization does not arise by writing the matrix as a product. Here the formula is extended to an operator on a Hilbert space, and a quick proof is given along the lines of earlier proofs.

Theorem 4.8 If k is the ascent of A, which has a Drazin inverse, then

$$A^D = A^k (A^{2k+1})^+ A^k.$$

Proof Since A^D exists, both the ascent and descent of A equal k. Because $A^\ell (A^\ell)^+$ is the orthogonal projection onto $A^\ell H$ for $\ell = 1,2,\ldots$, it follows that

$$A^k (A^k)^+ = A^{k+1} (A^{k+1})^+ = \ldots .$$

Induction and the formula $A^D = (A^D)^2 A$ show that $(A^D)^{\ell+1} A^\ell = A^D$ for $\ell = 1,2,\ldots$. It follows from these observations and the definition of A^D that

$$A^k (A^{2k+1})^+ A^k = (A^D)^{k+1} A^{2k+1} (A^{2k+1})^+ A^k$$

$$= (A^D)^{k+1} A^k (A^k)^+ A^k = (A^D)^{k+1} A^k = A^D . \quad \square$$

REFERENCES

1 F. V. Atkinson, "On relatively regular operators" , Acta. Sci. Math. (Szeged) 15 (1953), 38-56.

2 A. Ben-Israel and A. Charnes, "Contributions to the theory of generalized inverse", SIAM J. App. Math. 11 (1963), 667-699.

3 A. Ben-Israel and I. Erdelyi, "Extremal solutions of linear equations and generalized inversion between Hilbert spaces", J. Math. Anal. App. 39 (1972), 298-313.

4 A. Ben-Israel and T. N. E. Greville, "Generalized Inverses: Theory and Applications", John Wiley, New York, 1974.

5 R. H. Bouldin, "The product of operators with closed range", Tohoku Math. J. 25(1973), 359-363.

6 R. H. Bouldin, "The pseudo-inverse of a product", SIAM J. Appl. Math. 25(1973), 489-495.

7 R. H. Bouldin, "Closed range and relative regularity for products", J. Math. Anal. Appl., to appear.

8 S. L. Campbell, C. D. Meyer and N. J. Rose, "Applications of the Drazin inverse to linear systems of differential equations with singular coefficients", SIAM J. Appl. Math. 31 (1976), 411-425.

9 S. R. Caradus, "Operator Theory of the Pseudo-inverse", Queen's University notes, Kingston, Ontario, 1974.

10 R. E. Cline, "Inverses of rank invariant powers of a matrix", SIAM J. Numer. Anal. 5(1968) 182-197.

11 C. A. Desoer and B. H. Whalen, "A note on pseudoinverses, SIAM J. Appl. Math. 11 (1963), 442-447.

12 M. P. Drazin, "Pseudo-inverses in associative rings and semi-groups", Amer. Math. Monthly 65(1958), 506-514.

13 S. Goldberg, "Unbounded Operators, Theory and Applications", McGraw-Hill, New York, 1966.

14 C. W. Groetsch, "Computational theory of generalized inverses of bounded linear operators", University of Cincinnati notes, Cincinnati, 1976.

15 R. B. Holmes, "A Course on Optimization and Best Approximation", Springer-Verlag, New York, 1972.

16 T. Kato, "Perturbation Theory for Linear Operators", Springer-Verlag, New York, 1966.

17 J. J. Koliha, "The product of relatively regular operators", Commentationes Math. Univ. Carolinae 16(1975), 531-539.

18 D. C. Lay, "Spectral properties of generalized inverses of linear operators", SIAM J. Appl. Math. 29 (1975), 103-109.

19 J. L. Massera and J. J. Schaffer, "Linear Differential Equations and Function Spaces", Academic Press, New York, 1966.

20 M. Z. Nashed, "Generalized inverses, normal solvability, and iteration for singular operator equations", Nonlinear Functional Analysis and Applications, Academic Press, New York, 1971.

21 M. Z. Nashed and G. F. Votruba, "A unified operator theory of generalized inverses, Generalized Inverses and Applications, Academic Press, New York, 1976.

Richard Bouldin
University of Georgia
Athens, Georgia

Manuscript received Fall, 1976.

S L CAMPBELL

The Drazin inverse of an operator

1. INTRODUCTION

The Drazin inverse has proved helpful in analyzing Markov chains, difference equations, differential equations [3], [4], [5] and iterative procedures [14]. It would be useful if these results could be extended to infinite dimensional situations. Applications could then be made to denumerable Markov chains [12], abstract Cauchy problems [17], infinite systems of linear differential equations [13], [18], and possibly partial differential equations [9].

The difficulty lies in defining a Drazin inverse. One must first distinguish between an infinite matrix and a linear transformation on an infinite dimensional space. In working with denumerable Markov chains, infinite matrices are used and multiplication is not associative [12]. In the abstract Cauchy problem, that is $A\dot{x} + Bx = f$, the coefficients of the differential equation may be linear transformations or infinite matrices or both.

If T is a bounded linear transformation on a Banach space with spectrum, $\sigma(T)$, and 0 is a pole of order k of $(\lambda I - T)^{-1}$, then one can define

$$T^D = \frac{1}{2\pi i} \int_C \lambda^{-1}(\lambda I - T)^{-1} d\lambda,$$

and $k = \text{Index}(T)$. Here C is a contour around $\sigma(T)\backslash\{0\}$ with 0 on the outside of the contour. For operators of this type the usual theory goes through almost immediately since

$$T = P \begin{bmatrix} R & 0 \\ 0 & N \end{bmatrix} P^{-1}$$

with R invertible, $N^k = 0$, and P a bounded invertible linear transformation. This is essentially the approach of [7]. The papers [9], [10], [15], [17] do not explicitly mention the Drazin inverse but are concerned with resolvent integrals of $(\lambda A + B)^{-1}$ which are closely related [3]. While they differ on some points such as whether the operators are closed or bounded, all place assumptions on the growth of $(\lambda A + B)^{-1}$. For particular operators A,B this can be difficult information to determine.

One alternative approach is described in [2]. The basic idea there is to take a class $\mathcal{A}$ of operators, for example the commuting (2)-inverses of T, induce an order, and take maximal elements if they exist. This approach was used in [2] to define a group inverse for certain infinite matrices and extend some of the results of [5] to denumerable Markov chains.

In [2], $\mathcal{A}$ was defined as the set of all matrices X, with finite entries, such that $XA = AX$ and $X(AX) = (XA)X = X$. For $X,Y \in \mathcal{A}$, we defined $X \leq Y$ if $X(AY) = (XA)Y = X$, and $Y(AX) = (YA)X = X$. That is, A^D was a maximal commuting (2)-inverse. For $A \in \mathbb{C}^{n\times n}$ this is equivalent to the usual definition [2].

As discussed in [2], one difficulty with the operator case is in deciding what operators to include in $\mathcal{A}$. For example, does one want T^D to be bounded? How important is the commutivity of T and T^D? It is the author's feeling that the best approach is to define the class $\mathcal{A}$ with a particular application in mind.

This paper will discuss one such application and a type of Drazin inverse that is useful in analyzing it.

2. INFINITE LINEAR SYSTEMS

This section will consider the differential equation

$$A\dot{x} + Bx = f \qquad 0 \leq t \leq t_1 \tag{1}$$

where A,B are continuous linear operators from a Banach space $\mathcal{X}$ into a Banach space $\mathcal{Y}$. That is, $A,B \in \mathcal{B}(\mathcal{X},\mathcal{Y})$.

Let $D_{t_1}[A,B]$ be the set of all $x_o \in \mathcal{X}$ such that

$$A\dot{x} + Bx = 0, \tag{2}$$

with $x(0) = x_o$ has a solution on $[0,t_1]$ which is differentiable with respect to the norm of $\mathcal{X}$. The consistent initial conditions for (1) are a translate for those of (2) if (1) is consistent.

Drazin inverses are also useful in analyzing distributional or impulsive solutions [3], [4], [8] but we shall not do so in this paper. If $\mathcal{X} = \mathcal{Y}$, $B(\mathcal{X},\mathcal{Y})$ will be denoted $B(\mathcal{X})$.

$D_{t_1}[A,B]$ has been characterized for both finite dimensional [3] and certain infinite dimensional [17] systems using Laplace transforms.

The Theorems of [3], [5] provide a nice geometric characterization of $D_{t_1}[A,B]$ in the finite dimensional case if there exists a λ such that $\lambda A + B$ has a right or left inverse. We shall establish infinite dimensional versions of these characterizations, then give an example showing the difficulty of further extension. Drazin and weak Drazin inverses for a larger class than those covered by [7] will be defined.

For any operator T let $\mathcal{R}(T)$ be its range and $\mathcal{N}(T)$ be its null space. Let

$$\mathcal{R}(T^{\infty}) = \bigcap_{n=0}^{\infty} \mathcal{R}(T^{n}).$$

A subspace is a closed linear manifold. The restriction of an operator T to an invariant subspace $\mathcal{M}$ is denoted $T|\mathcal{M}$. To avoid technical problems we shall consider bounded operators. Our approach can be used to study unbounded operators also.

<u>Theorem 1</u> Suppose that $A,B \in B(\mathcal{X},\mathcal{Y})$, and there exists a λ such that $\lambda A + B$ is invertible. Let $\hat{A}_{\lambda} = (\lambda A + B)^{-1}A$. Suppose that $\mathcal{R}(\hat{A}_{\lambda}^{k})$ is closed for all $k \geq 0$. Then

$$D_{t_1}[A,B] \subseteq \mathcal{R}(\hat{A}_{\lambda}^{\infty}) \qquad \text{for all } 0 < t_1 \leq \infty. \tag{3}$$

If $\mathcal{N}(\hat{A}_{\lambda}) \cap \mathcal{R}(\hat{A}_{\lambda}^{\infty})$ is complemented in $\mathcal{R}(\hat{A}_{\lambda}^{\infty})$, then

$$D_{t_1}[A,B] = \mathcal{R}(\hat{A}_{\lambda}^{\infty}) \qquad \text{for all } 0 < t_1 \leq \infty . \tag{4}$$

<u>Proof</u> Note that x is a solution of (2) if and only if

$$\hat{A}_{\lambda}\dot{x} + (I - \lambda\hat{A}_{\lambda})x = 0. \tag{5}$$

From (5), $x(t) \in \mathcal{R}(\hat{A}_{\lambda})$ for all t. Hence $\dot{x}(t) \in \mathcal{R}(\hat{A}_{\lambda})$ for all t since $\mathcal{R}(\hat{A}_{\lambda})$ is closed. But then $x(t) \in \mathcal{R}(\hat{A}_{\lambda}^{2})$ for all t. Proceeding we get

$x(t) \in R(\hat{A}_\lambda^\infty)$ for all t, and (3) follows. Now $\hat{A}_\lambda$ maps $R(\hat{A}_\lambda^\infty)$ onto itself. If $N(\hat{A}_\lambda) \cap R(\hat{A}_\lambda^\infty)$ is complemented in $R(\hat{A}_\lambda^\infty)$, then $\hat{A}_\lambda$ restricted to $R(\hat{A}^\infty)$ has a bounded right inverse C. Thus for any $x_o \in R(\hat{A}_\lambda^\infty)$, $x(t) = e^{-C(I-\lambda\hat{A}_\lambda)t} x_o$ is a solution of (5). Hence (4) follows. □

Of course, in a Hilbert space, the complementation assumption is not needed. The assumptions on $R(\hat{A}_\lambda^k)$ are needed in some form.

Example 1 Take $X = \ell^2$, $A = \mathrm{Diag}\{1/n\}$, $B = -I$. Note that all assumptions of Theorem 1 are met except for $R(\hat{A}_\lambda^k)$ being closed. Clearly $R(\hat{A}_\lambda^\infty) = R(A^\infty)$. (3) may be rewritten as $\dot{x}_n - nx_n = 0$, $n \geq 1$. Thus

$x_n(t) = e^{nt}x_n(0)$. Let $y_n = e^{-n}$. Then $\{y_n\} \in R(A^\infty)$ since $\{n^k y_n\} \in \ell^2$ for all $k \geq 0$. But $\{x_n(t)\} = \{e^{nt}e^{-n}\} \notin \ell^2$ for all $t > 0$. Thus $\{y_n\} \in R(A^\infty)$, but $\{y_n\} \notin D_{t_1}[A,B]$ for any $t_1 > 0$, so that (4) does not hold. Note that in this case, (3) does.

This example also shows that $D_{t_1}[A,B]$ can vary with t_1 since $D_{t_1}[A,B] \supsetneq D_{t_2}[A,B]$ if $t_2 > t_1$.

Example 2 Again take $X = \ell^2$, A as in Example 1, but B = I. Then $R(\hat{A}^\infty)$ is the same as before. But for any $\{y_n\} \in A\ell^2$, $\{e^{-nt}y_n\}$ satisfies (3). Thus there exists $\{y_n\} \notin R(A^\infty)$ such that $\{y_n\} \in D_{t_1}[A,B]$ for all $t_1 > 0$.

Examples 1 and 2 show that as soon as one removes the assumption of closed ranges, it becomes necessary to consider spectral properties of $\hat{A}_\lambda$ and the pencil $\lambda\hat{A} + \hat{B}$ as done in [17]. If in Theorem 1, $\lambda A + B$ has a bounded left inverse and $R(\hat{A}_\lambda^k)$ is closed for all k, then (3) still holds.

In the finite-dimensional case, $R(\hat{A}_\lambda^\infty)$ is independent of λ. The same holds for the case covered by Theorem 1.

Theorem 2 Suppose $A,B \in B(X,Y)$ and λ,μ are such that $\lambda A + B$, $\mu A + B$ are invertible. Then $R(\hat{A}_\lambda^\infty) = R(\hat{A}_\mu^\infty)$.

Proof $\hat{A}_\lambda = (\lambda A + B)^{-1}A + (\lambda\hat{A}_\mu + (I - \mu\hat{A}_\mu))^{-1}(\mu A + B)^{-1}A$

$= (I + (\lambda-\mu)\hat{A}_\mu)^{-1}\hat{A}_\mu$. Thus $R(\hat{A}_\lambda^k) = R(\hat{A}_\mu^k)$ since $\hat{A}_\mu$ commutes with the invertible operator $(I + (\lambda-\mu)\hat{A}_\mu)$. □

With applications to (1) in mind, and motivated by the proof of Theorem 1 a Drazin-like inverse will now be defined.

Definition 1 If $A \in B(X)$, $R(A^\infty)$ is closed, and $R(A^\infty)$ is complemented in X, then a special type of weak Drazin inverse [6] is defined as follows. Let M be a complementary subspace to $R(A^\infty)$. Relative to $R(A^\infty) \oplus M$, A may be written

$$A = \begin{bmatrix} U & S \\ 0 & V \end{bmatrix} \tag{6}$$

where U is an onto bounded operator, and $R(V^\infty) = \{0\}$. If U has a bounded right inverse, then define

$$A^d = \begin{bmatrix} U^o & U^oZ \\ 0 & 0 \end{bmatrix},$$

where U^o is a bounded right inverse of U, Z is an arbitrary bounded operator.

Note that A^d can exist for operators for which 0 is not only not an isolated point of the spectrum of A but in fact, an interior point of the spectrum.

Definition 2 If in (6) U is invertible and $S = 0$, then A^d is called the Drazin inverse of A and is denoted A^D.

Note that different M's may give $S = 0$ so A^D need not be unique. (See Example 3).

Proposition 1 Suppose $A \in B(X)$, and $R(A^\infty)$ is closed and complemented by a subspace M. Let P be the projection onto M along $R(A^\infty)$. If $\sigma(A|R(A^\infty)) \cap \sigma(PA \mid M) = \emptyset$, then A^D exists.

Proof The assumptions of Proposition 1 imply $\sigma(U) \cap \sigma(V) = \emptyset$ in (6). Since $0 \in \sigma(V)$, U is invertible and A is similar to $\begin{bmatrix} U & 0 \\ 0 & V \end{bmatrix}$ [16]. □

Note that if the index of A is finite, then V in (6) is nilpotent of index k and Definition 2 gives the usual Drazin inverse. If A^d exists and V is not nilpotent, A will be said to be of infinite index.

Proposition 2 Suppose $A \in B(X)$, and A^D exists. Then

(i) A^D is unique under the assumptions of Proposition 1

(ii) $A^D A = AA^D$

(iii) $R(A^D A) = R(A^\infty)$

(iv) $A^D AA^D = A^D$.

If A^d exists, then

(v) $R(AA^d) = R(A^\infty)$,

(vi) $A^d AA^d = A^d$.

If U in (6) is invertible, then

(vii) $R(A^d A) = R(A^\infty)$.

The solution of (2) is given by;

Corollary 1 If $A,B \in B(X,Y)$, there exists a λ such that $\lambda A + B$ is invertible, $R(\hat{A}_\lambda^{\,k})$ is closed for all k, $N(\hat{A}_\lambda) \cap R(\hat{A}_\lambda^{\,\infty})$ is complemented in $R(\hat{A}_\lambda^{\,\infty})$, and $R(\hat{A}_\lambda^{\,\infty})$ is complemented in X, then

$$x(t) = e^{-\hat{A}_\lambda^{\,d}(I-\lambda\hat{A}_\lambda)t}\hat{A}_\lambda\hat{A}_\lambda^{\,d}q \tag{7}$$

is a solution of (2) for all $q \in X$. Also $R(\hat{A}_\lambda\hat{A}_\lambda^{\,d}) = D_{t_1}[A,B]$. Solutions are uniquely determined by consistent initial conditions if and only if

$N(\hat{A}_\lambda) \cap R(\hat{A}_\lambda^\infty) = \{0\}$. In this case (7) gives all solutions of (2).

In solving (1), the appropriate formulas could be developed under the weaker assumption that $\hat{A}_\lambda^d$ exists. For simplicity of exposition we shall only consider the case when $\hat{A}_\lambda^D$ exists. For notational convenience the lambda subscript will be omitted from $\hat{A}_\lambda$.

Proposition 3 Suppose that A,B satisfy the assumption of Corollary 1 and $\hat{A}^D$ exists. Then all solutions of (1) are of the form

$$x = e^{-\hat{A}^D(I-\lambda\hat{A})t}\hat{A}^D\hat{A}q + \hat{A}^D \int_0^t e^{\hat{A}^D(I-\lambda\hat{A})(s-t)}\hat{f}(s)ds + (I - \hat{A}^D\hat{A})x(t) \qquad (8)$$

where $(I - \hat{A}^D\hat{A})x(t)$ is the unique solution, if it exists, of

$$\hat{A}[(I - \hat{A}^D\hat{A})\dot{x}] + (I - \lambda\hat{A})\,[(I- \hat{A}^D\hat{A})x] = (I - \hat{A}^D\hat{A})f. \qquad (9)$$

In [3], [5] the explicit solution of (9) was given as

$$(I - \hat{A}^D\hat{A})\,x = \sum_{i=0}^{k-1} [-\hat{A}^D(I - \lambda\hat{A})]^i(I - \lambda\hat{A})^D\hat{f}^{(i)} \qquad (10)$$

when X was finite dimensional. If $\mathrm{Index}(\hat{A}(I - \hat{A}^D\hat{A})) < \infty$ and X is a Hilbert space, then $\hat{A}(I - \hat{A}^D\hat{A})$ has the usual Jordan form since the subspaces $R(\hat{A}^k)$ are assumed closed [11]. In general, one has $\hat{A}(I - \hat{A}^D\hat{A})$ has a strictly lower triangular block matrix representation and the entries of $(I - \hat{A}^D\hat{A})x$ with respect to this decomposition may be computed recursively in a unique manner. The difficulty is in showing that $(I - \hat{A}^D\hat{A})x$ when so calculated is in X.

For purposes of illustration consider the case when (9) takes the form

$$\begin{bmatrix} 0 & 0 & 0 \\ V_1 & 0 & 0 \\ 0 & V_2 & 0 \\ \cdot & \cdot & \cdot \end{bmatrix} \begin{bmatrix} \dot{x}_1 \\ \dot{x}_2 \\ \dot{x}_3 \\ \cdot \end{bmatrix} + \begin{bmatrix} I & 0 & 0 \\ 0 & I & 0 \\ 0 & 0 & I \\ \cdot & \cdot & \cdot \end{bmatrix} \begin{bmatrix} x_1 \\ x_2 \\ x_3 \\ \cdot \end{bmatrix} = \begin{bmatrix} f_1 \\ f_2 \\ f_3 \\ \cdot \end{bmatrix}, \qquad (11)$$

Then $x_1 = f_1$ and $V_i\dot{x}_i + x_{i+1} = f_{i+1}$. This would occur, for example, if V were similar to a block weighted shift and $I - \lambda V$ were invertible. Then the solution of (11) is

$$x_i = \sum_{k=0}^{i} (-1)^k \prod_{j=i-1-k}^{i-1} V_j f_{i-k}^{(k)}(t). \tag{12}$$

Thus for (11) to be consistent, the f_i must be infinitely differentiable. The expression (12) may be written as

$$x = \sum_{n=0}^{\infty} (-1)^n [V(I - \lambda V)^{-1})]^n (I - \lambda V)^{-1} (I - \hat{A}^D \hat{A}) f^{(i)} \tag{13}$$

which is a direct analogue of (10). Note (11) need not be consistent for all f. For example, if $V_i = I$, $f_i = 0$ for $i > 1$, $f_1(t) = e^t$, $x(t)$ would be in ℓ^∞ but not in ℓ^2.

At this point, it should be clear that one can use Proposition 3 to get an analogue of (10) where the f's are restricted to those for which (1) is consistent except that the sum will be infinite as in (13).

We conclude this paper by noting that our A^D could have also been defined as in [2] as follows.

Proposition 4 Suppose that A satisfies the assumptions of Definitions 1, 2 and Proposition 1. Let $\mathcal{A} = \{Y \in B(\mathcal{X}) \mid YA = AY, YAY = Y\}$ and order $\mathcal{A}$ by $Y \leq Z$ if $YAZ = ZAY = Y$. Then A^D is the unique maximal element of $\mathcal{A}$.

Proof It suffices to show that $A^D \in \mathcal{A}$ and is maximal. Suppose $Y \in \mathcal{A}$. Then $\mathcal{R}(A^\infty)$ is invariant under Y so that subject to the appropriate decomposition of $\mathcal{X}$,

$$Y = \begin{bmatrix} Y_1 & Y_2 \\ 0 & Y_3 \end{bmatrix}, \quad A = \begin{bmatrix} U & 0 \\ 0 & V \end{bmatrix}$$

Then $Y_3V = VY_3$, and $Y_3VY_3 = Y_3$. Now Y_3V is a projection and $Y_3V = (Y_3V)^n = V^nY_3{}^n$ so that $\mathcal{R}(Y_3V) \subseteq \mathcal{R}(V^n)$. Hence $\mathcal{R}(Y_3V) = \{0\}$ and $Y_3V = 0$. Thus $Y_3 = 0$. Since $A^DA = \begin{bmatrix} I & 0 \\ 0 & 0 \end{bmatrix} = A A^D$, $Y \leq A^D$ if $Y_2 = 0$.

That $Y_2 = 0$ follows from the assumption that $\sigma(A|\mathcal{R}(A^\infty)) \cap \sigma(PA|\mathcal{M}) = \emptyset$. If this condition does not hold, then it is sometimes possible to construct

$Y \in \mathcal{A}$ such that $Y_1 = U^{-1}$, $Y_2 \neq 0$ and neither $Y \leq A^D$ nor $A^D \leq Y$ holds.

Example 3 Let $X = M \oplus \ell^2$ be the Hilbert space of square summable sequence $\{a_n\}_{n=-\infty}^{\infty}$, where the decomposition $M \oplus \ell^2$ is $\{a_n\}_{n=-\infty}^{-1} \oplus \{a_n\}_{n=0}^{\infty}$. Let U be the shift operator (bilateral) on X. Let $V = 0 \oplus S$ where S is the unilateral shift on ℓ^2. Define

$$Y = \begin{bmatrix} U^{-1} & V \\ 0 & 0 \end{bmatrix}, \quad A = \begin{bmatrix} U & 0 \\ 0 & V \end{bmatrix}; \quad A^D = \begin{bmatrix} U^{-1} & 0 \\ 0 & 0 \end{bmatrix}$$

so that $Y \in \mathcal{A}$. Note that $YAA^D = A^D$ while $A^D AY = Y$ so that A^D and Y are not comparable. Also note that

$$\begin{bmatrix} I & V \\ 0 & I \end{bmatrix} \begin{bmatrix} U & 0 \\ 0 & V \end{bmatrix} \begin{bmatrix} I & V \\ 0 & I \end{bmatrix} = \begin{bmatrix} U & 0 \\ 0 & V \end{bmatrix}$$

but

$$\begin{bmatrix} I & V \\ 0 & I \end{bmatrix} \begin{bmatrix} U^{-1} & 0 \\ 0 & 0 \end{bmatrix} \begin{bmatrix} I & V \\ 0 & I \end{bmatrix} \begin{bmatrix} U^{-1} & -U^{-1}V \\ 0 & 0 \end{bmatrix} \neq A^D.$$

In this example the Drazin inverse is not unique. The set of Drazin inverses is preserved by similarity.

REFERENCES

1 R. Bouldin, Generalized inverses and factorizations, this book.

2 S. L. Campbell, The Drazin inverse of an infinite matrix, SIAM J. Appl. Math., 31(1976), 492-503.

3 S. L. Campbell, Singular Systems of Differential Equations, Pitman, 1980.

4 S. L. Campbell, Singular Systems of Differential Equations, II, Pitman, 1982, (in press).

5 S. L. Campbell and C. D. Meyer, Jr., Generalized Inverses of Linear Transformations, Pitman Publishing Co., Ltd., 1979.

6 S. L. Campbell, Weak Drazin inverses, Linear Alg. and Its Appl., 20(1978), 167-178.

7 S. R. Caradus, Operator Theory of the Pseudo-inverse, Queen's University Notes, Kingston, Ontario, 1974.

8 J. D. Cobb, Descriptor Variable and Generalized Singularly Perturbed Systems: a Geometric Approach, Ph.D. Thesis, Department of Electrical Engineering, University of Illinois, 1980.

9 A. Favini, Abstract potential operators and spectral methods for a class of degenerate evolution problems, J. Diff. Eqn., 39, 1981, 212-225.

10 A. Favini, Laplace transform method for a class of degenerate evolution problems, Rend. Mat., 12(1979), 511-536.

11 L. J. Gray, Jordan representation for a class of nilpotent operators, Ind. Univ. Math. J., 26(1977), 57-64.

12 J. G. Kemeny, J. L. Snell, and A. W. Knapp, Denumerable Markov Chains, Van Nostrand, Princeton, N. J., 1966.

13 J. P. McClure and R. Wong, On infinite systems of linear differential equations, Can. J. Math. XXVII, (1975), 691-703.

14 C. D. Meyer, Jr. and R. J. Plemmons, Convergent powers of a matrix with applications to iterative methods for singular linear systems, SIAM J. Numer. Anal., 14(1977).

15 N. I. Radbel, Initial manifold and the dissipativity of a Cauchy problem for the equation $A\dot{x}(t) + Bx(t) = 0$, Diff. Eqn. (translation of Diff. Urav.), 15(1979), 810-811.

16 M. Rosenblum, On the operator equation $BX - XA = 0$, Duke Math. J., 23(1956), 263-270.

17 A. G. Rutkas, Cauchy's Problem for the equation, $Ax'(t) + Bx(t) = f(t)$, Diff. Eqn. (translation of Diff. Urav.), 11(1975), 1486-1497.

18 S. Steinberg, Infinite systems of ordinary differential equations with unbounded coefficients and moment problems, J. Math. Anal. Appl., 41(1973), 685-694.

Stephen L. Campbell
Department of Mathematics
North Carolina State University
Raleigh, North Carolina 27650

Research sponsored by the Air Force Office of Scientific Research, Air Force Systems Command, USAF, under Grant No. AFOSR-81-0052. The United States Government is authorized to reproduce and distribute reprints for Governmental purposes notwithstanding any copyright notation hereon.

I ERDELYI

Spectral decompositions for generalized inversions

NOTATIONS

$\mathbb{C}$, the complex field.

$\underline{G}$, the family of all open sets in $\mathbb{C}$.

For a set $A \subset \mathbb{C}$, we denote by

$\bar{A}$, the closure;

A^c, the complement (with respect to a given total set);

∂A, the boundary.

For a linear operator T on a Banach space X, we denote by

D_T, the domain;

R_T, the range;

Ker T, the kernel (the null manifold);

C_T, the carrier = $D_T \cap (\text{Ker } T)^{\perp}$;

T', the conjugate operator;

T*, The Hilbert space adjoint;

$T^{\dagger}$, the generalized inverse;

$\sigma(T)$, the spectrum;

$\sigma_T(x)$, the local spectrum;

$\rho(T)$, the resolvent set;

$\rho_T(x)$, the local resolvent set;

$R(\cdot,T)$, the resolvent operator;

$\tilde{x}(\cdot)$, the local resolvent operator.

For a member Y of the family Inv(T) of all invariant subspaces (closed linear manifolds) under T, we denote by

$T|Y$, the restriction of T to Y;

T/Y, the coinduced operator on the quotient space X/Y.

For linear manifolds V, W in X, we denote by

$V^{\perp}$, the annihilator of V, i.e. $V^{\perp} = \{x' \in X' : \langle x,x' \rangle = 0 \text{ for all } x \in V\}$;

$V \oplus^{\perp} W$, the orthogonal sum in a Hilbert space,

$V \vee W$, the smallest subspace that contains $V \cup W$.

$V \oplus W$, the direct (topological) sum (also used for operators);

$\lambda - T$, for $\lambda I - T$ (I, the identity operator);

λ-T|Y, for $\lambda I|Y - T|Y$;

λ-T/Y, for $\lambda I/Y - T/Y$;

SDP, spectral decomposition property;

SVEP, single valued extension property.

INTRODUCTION

The generalized inverse, defined for bounded linear operators between Banach spaces and for a wide variety of linear operators between Hilbert spaces, entails direct sum decompositions of certain linear manifolds in the underlying spaces.

The existence of Tseng's maximal generalized inverse [19] between Hilbert spaces is conditioned by the decomposable domain [2]

$$D_T = C_T \oplus^{\perp} \operatorname{Ker} T$$

of the given operator T. If T does not have a decomposable domain, then a broader type of generalized inverse $T^{\dagger}$ has such a domain [10]

$$D_{T^{\dagger}} = T(C_T) \oplus^{\perp} [T(C_T)]^{\perp}$$

or [11]

$$D_{T^{\dagger}} = T(C_T) \oplus^{\perp} (R_T)^{\perp} .$$

For an operator-theoretic description of the generalized inversion between Hilbert spaces, the reader is directed to Chapter 8 of monograph [6].

A bounded linear operator T between two Banach spaces X, Y has a generalized inverse $T^{\dagger}$ verifying equations

$$TT^{\dagger}T = T, \quad T^{\dagger}TT^{\dagger} = T^{\dagger} \tag{1}$$

iff Ker T and R_T are closed and have topological complements in X and Y, respectively. In this case, $TT^{\dagger}$ is a bounded projection of Y onto R_T and $T^{\dagger}T$ is a bounded projection of X onto $R_{T^{\dagger}}$. Thus T in (1) has complemented range and kernel (see e.g. [3, Section 5]).

More generally [4], two subspaces Y and Z of a Banach space X are said to be quasi-complements in X if Y + Z is dense in X. Then there exists a closed linear operator P with domain

$$D_P = Y + Z$$

and defined by

$$P(y + z) = y \quad \text{with } y \in Y \text{ and } z \in Z.$$

Thus P has range Y and kernel Z. Moreover, for every $x \in D_p$,

$$Px \in D_p \quad \text{and } P^2x = Px. \tag{2}$$

Conversely, if P is a closed linear operator with domain D_p dense in X and satisfying condition (2), then the range R_p and the kernel Ker P are quasi-complements in X and

$$D_p = R_p + \text{Ker } P.$$

A densely defined closed linear operator P satisfying condition (2) is called a quasi-projection in X. Thus, there is a one-to-one correspondence between pairs of quasi-complementary subspaces of X and quasi-projections in X. The spectral-theoretic framework for quasi-complementations in a Banach space is the asymptotic spectral decomposition [13, Chapter III]. This can be applied to generalized inversions of densely defined closed operators on a Banach space.

We shall confine the present considerations to bounded linear operators of a Banach algebra B(X) acting on a Banach space X.

For $T \in B(X)$, Dunford's formula (e.g. [8, VII. 3.9])

$$f(T) = \frac{1}{2\pi i} \int_{\Gamma} f(\lambda)R(\lambda;T)d\lambda$$

where Γ is an admissible contour which surrounds $\sigma(T)$ and is contained in $\rho(T)$, establishes the isomorphic mapping $f \to f(T)$ of the functional calculus. If $\sigma(T)$ is disconnected, then

$$E = \frac{1}{2\pi i} \int_{\Gamma} R(\lambda;T)d\lambda$$

is a projection whenever the admissible contour Γ "separates" the disconnected parts of the spectrum.

In this study we shall use some techniques of the contemporary spectral theory to produce direct sum decompositions needed for generalized inversions on an abstract Banach space X. We shall explore some cases of a disconnected spectrum $\sigma(T)$ for a given $T \in B(X)$.

1. ELEMENTS OF SPECTRAL THEORY

Definition 1.1 A spectral decomposition of X by T is a finite system $\{(G_i,Y_i)\}_1^n \subset \underline{G} \times \text{Inv}(T)$ with the following properties:

(i) $\sigma(T) \subset \bigcup_{i=1}^{n} G_i$;

(ii) $X = \sum_{i=1}^{n} Y_i$;

(iii) $\sigma(T|Y_i) \subset G_i$, $i = 1,2,\ldots,n$.

Definition 1.2 [12] T is said to have the spectral decomposition property (abbreviated SDP) if for any finite open cover $\{G_i\}_1^n$ of $\sigma(T)$, there is a system $\{Y_i\}_1^n \subset \text{Inv}(T)$ such that $\{(G_i,Y_i)\}_1^n$ is a spectral decomposition of X by T.

The spectra $\sigma(T|Y_i)$ may contain some bounded components of $\sigma(T)$, known as holes in the spectrum of T. Therefore, the spectral inclusion property $\sigma(T|Y_i) \subset \sigma(T)$ is not always satisfied by the invariant subspaces Y_i of the spectral decomposition. If for an invariant subspace Y under T, we have $\sigma(T|Y) \subset \sigma(T)$, then Y is said to be a ν-space [5]. A necessary and sufficient condition for an invariant subspace Y to be a ν-space is expressed by the following:

Proposition 1.3 [18] Given $T \in B(X)$, an invariant subspace Y is a ν-space iff Y is invariant under the resolvent operator, i.e. iff $R(\lambda;T)Y \subset Y$ for all $\lambda \in \rho(T)$.

Analytic continuations are instrumental in the spectral theory. Dunford [7] introduced the concept of the single valued extension property by the following:

Definition 1.4 $T \in B(X)$ is said to have the single valued extension property (abbreviated SVEP) if for any function $f : D\ (\subset \mathbb{C}) \to X$ analytic on an open set D, the condition

$$(\lambda-T)f(\lambda) = 0 \quad \text{on } D$$

implies $f = 0$.

Equivalently, for every $x \in X$, any two analytic extensions f and g on $R(\lambda;T)x$ agree on $D_f \cap D_g$. When this property holds, then the union of all analytic extensions of $R(\lambda;T)x$ is called the local resolvent set and is denoted by $\rho_T(x)$. The SVEP implies the existence of a maximal analytic extension $\tilde{x}(\cdot)$ of $R(\cdot;T)x$ to $\rho_T(x)$, called the local resolvent operator. This function identically verifies the equation

$$(\lambda-T)\tilde{x}(\lambda) = x \quad \text{on } \rho_T(x).$$

The local spectrum $\sigma_T(x)$, defined as the complement in $\mathbb{C}$ of $\rho_T(x)$, is the set of the singularities of $\tilde{x}$. By definition

$$\sigma_T(x) \subset \sigma(T) \quad \text{for all } x \in X.$$

Some immediate properties of the local spectrum are included in the following

Proposition 1.5 [8] Given $T \in B(X)$ with the SVEP, the following properties hold:

(i) $\sigma_T(x + y) \subset \sigma_T(x) \cup \sigma_T(y)$ for $x,y \in X$;

(ii) $\sigma_T(x) = \emptyset$ iff $x = 0$;

(iii) $\sigma_T(Sx) \subset \sigma_T(x)$ for every $S \in B(X)$ which commutes with T;

(iv) $\sigma_T(y) \subset \sigma_{T|Y}(y)$, for every $Y \in \mathrm{Inv}(T)$ and all $y \in Y$.

Proof Since properties (i) - (iii) are well-known (e.g.[8, XVI]), we shall prove (iv). It is clear that the SVEP is inherited by the restriction of T to any invariant subspace Y. For every $y \in Y$ and $\lambda \in \rho_{T|Y}(y)$, we have

$$(\lambda-T|Y)y(\lambda) = (\lambda-T)y(\lambda) = y$$

and hence $\rho_{T|Y}(y) \subset \rho_T(y)$. □

Corollary 1.6 If $T \in B(X)$ has the SVEP, then for any subset H of C, the set

$$X_T(H) = \{x \in X : \sigma_T(x) \subset H\} \tag{1.1}$$

is a linear manifold in X and has the following properties; $X_T(\emptyset) = \{0\}$, $X_T(H \cap K) = X_T(H) \cap X_T(K)$, and $H \subset K$ implies $X_T(H) \subset X_T(K)$.

Proof is an immediate consequence of (1.1) and Proposition 1.5. □

It follows from (1.1), that for any $Y \in \mathrm{Inv}(T)$,

$$Y \subset X_T[\sigma(T|Y)]. \tag{1.2}$$

We shall call a subspace Y hyperinvariant under T if Y is invariant under every operator in B(X) which commutes with T.

Proposition 1.7 Let $T \in B(X)$ have the SVEP. For every subset H of $\mathbb{C}$, the subspaces $\overline{X_T(H)}$ and $X_T(H)^{\perp}$ are hyperinvariant under T and T', respectively.

Proof First, let $x \in \overline{X_T(H)}$ be given. There is a sequence $\{x_n\} \subset X_T(H)$ which converges (in the norm topology of X) to x. If $S \in B(X)$ commutes with T, then Proposition 1.5 (iii) implies that $\sigma_T(Sx_n) \subset \sigma_T(x_n) \subset H$ for all n. Then, for every n, $Sx_n \in X_T(H)$ and the continuity of S implies that $Sx \in \overline{X_T(H)}$. Thus, $S[\overline{X_T(H)}] \subset \overline{X_T(H)}$.

Next, let $y' \in X_T(H)^{\perp}$. By the first part of the proof, for every $x \in X_T(H)$ we have $\langle Sx, y' \rangle = 0$. Consequently,

$$0 = \langle Sx, y' \rangle = \langle x, S'y' \rangle \text{ for all } x \in X_T(H)$$

and hence $S'y' \in X_T(H)^{\perp}$. Thus $S'[X_T(H)^{\perp}] \subset X_T(H)^{\perp}$. □

The first important property of the operators with the SDP is expressed by the following

Theorem 1.8 [12] Every operator with the SDP has the SVEP.

The fact that the operators with the SDP have a well-developed spectral theory, will be seen in connection with the class of decomposable operators. For this we need some preparation.

Definition 1.9 [15] Given $T \in B(X)$, $Y \in \mathrm{Inv}(T)$ is said to be a spectral maximal space for T if for any $Z \in \mathrm{Inv}(T)$,

$$\sigma(T|Z) \subset \sigma(T|Y) \text{ implies } Z \subset Y.$$

It follows easily from the definition that every spectral maximal space is a ν-space. Another useful property of spectral maximal spaces now follows.

Lemma 1.10 Given $T \in B(X)$, let Y be a spectral maximal space for T. Then

$$x \in X, \quad \lambda \in \sigma(T|Y) \quad \text{and} \quad (\lambda - T)x \in Y \tag{1.3}$$

imply that $x \in Y$.

Proof Clearly, the hypotheses (1.3) imply the following inclusions

$$\sigma[T|(Y \vee \{x\})] \subset \sigma(T|Y) \cup \{\lambda\} \subset \sigma(T|Y).$$

Since Y is spectral maximal, we have $Y \vee \{x\} \subset Y$ and hence $x \in Y$. □

Definition 1.11 [15] $T \in B(X)$ is called decomposable if for every finite open cover $\{G_i\}_1^n$ of $\sigma(T)$, there is a system of spectral maximal spaces $\{Y_i\}_1^n$ such that $\{(G_i, Y_i)\}_1^n$ is a spectral decomposition of X by T.

If T is decomposable, then for every closed set $F \subset \mathbb{C}$, $X_T(F)$ is closed and it is a spectral maximal space for T. Moreover, if F is closed and T is decomposable, then [15]

$$\sigma[T|X_T(F)] \subset F. \tag{1.4}$$

Conversely, every spectral maximal space Y for a decomposable operator T has the representation [15]

$$Y = X_T[\sigma(T|Y)]. \tag{1.5}$$

The decomposable operators possess a remarkable duality property.

Theorem 1.12 [16] If T is decomposable on X, then the adjoint T' is decomposable on the dual space X'.

With regard to the coinduced operator T/Y for a decomposable T and a spectral maximal space Y, we have the following useful property.

Theorem 1.13 If T is decomposable, then for every open set $G \subset \mathbb{C}$, we have

$$\sigma[T/X_T(\overline{G})] \subset G^c.$$

Proof We shall show that for every $\lambda \in G$, the operator-valued function $\lambda - T/X_T(\overline{G}) : G \to X/X_T(\overline{G})$ is bijective.

For the proof of the injectivity, assume that $(\lambda - T)x \in X_T(\overline{G})$ for some $\lambda \in G$. If $\lambda \in \rho[T|X_T(\overline{G})]$, then

$$x = [\lambda - T|X_T(\overline{G})]^{-1}(\lambda - T)x \in X_T(\overline{G})$$

and if $\lambda \in \sigma[T|X_T(\overline{G})]$, then by Lemma 1.10, $x \in X_T(\overline{G})$. Thus, $\lambda - T/X_T(\overline{G})$

is injective on G.

To prove the surjectivity, let $\lambda \in G$ be arbitrary and choose an open set H such that $\sigma(T) \subset G \cup H$ and $\lambda \in \overline{H}^c$. We have the spectral decomposition

$$X = X_T(\overline{G}) + X_T(\overline{H})$$

with

$$\lambda \in \rho[T|X_T(\overline{H})]. \tag{1.7}$$

An arbitrary vector $x \in X$ has a representation

$$x = x_1 + x_2 \quad \text{with} \quad x_1 \in X_T(\overline{G}), \quad x_2 \in X_T(\overline{H}).$$

In view of (1.7), there is a vector $y \in X_T(\overline{H})$ such that $(\lambda-T)y = x_2$. Then the coset $\hat{y} = y + X_T(\overline{G})$ verifies equation

$$[\lambda - T/X_T(\overline{G})]\hat{y} = \hat{x}_2 = \hat{x}.$$

Thus $\lambda - T/X_T(\overline{G})$ is surjective and by the first part of the proof, bijective. Therefore, $\lambda \in \rho[T/X_T(\overline{G})]$ and since λ is arbitrary in G, $G \subset \rho[T/X_T(\overline{G})]$. □

Corollary 1.14 If T is decomposable, then for any open set $G \subset \mathbb{C}$, we have

$$\sigma[T'|X_T(G)^{\perp}] \subset G^c. \tag{1.8}$$

Proof Since the dual of the quotient space $X/X_T(\overline{G})$ is isometrically isomorphic to $X_T(G)^{\perp}$, (1.6) (Theorem 1.13) implies

$$\sigma[T'|X_T(G)^{\perp}] = \sigma([T/X_T(\overline{G})]') = \sigma[T/X_T(\overline{G})] \subset G^c. \quad \square$$

It has been recently discovered that the class of operators with the SDP coincides with that of decomposable operators.

Theorem 1.15 Every operator $T \in B(X)$ which has the SDP is decomposable on X.

This theorem was proved independently by E. Albrecht [1] and B. Nagy [17].

Theorem 1.15 makes the general spectral decomposition of a linear operator, given by Definition 1.2, a powerful tool for producing direct sum decompositions on both a given Banach space X and on its dual space X'.

2. DIRECT SUM DECOMPOSITIONS

We have mentioned in the Introduction that a disconnected spectrum produces a direct sum decomposition of the space $X = Y_1 \oplus Y_2$ and subsequently of the given $T \in B(X)$, $T = T_1 \oplus T_2$ with $T_1 = T|Y_1$ and $T_2 = T|Y_2$. We shall apply the general spectral decomposition theory to operators with disconnected spectra to obtain some more information on the invertibility or non-invertibility of the direct summands T_1 and T_2, to localize their spectra $\sigma(T_1)$ and $\sigma(T_2)$ in some given open sets, and to contrive corresponding direct sum decompositions on the dual space X'.

We begin with a special case.

Theorem 2.1 Let $T \in B(X)$ have the SDP and assume that 0 is an isolated point of the spectrum. Then T is the direct sum of an invertible and a quasinilpotent operator.

Proof Since 0 is an isolated point of $\sigma(T)$, there is a positive integer n such that

$$\{\lambda \in \mathbb{C} : 0 < |\lambda| < \tfrac{1}{n}\} \subset \rho(T).$$

Consider the open cover of $\sigma(T)$ formed by the sets

$$G_1 = \{\lambda \in \mathbb{C} : |\lambda| > \frac{1}{n+1}\},\ G_2 = \{\lambda \in \mathbb{C} : |\lambda| < \frac{1}{n+1}\}.$$

By the SDP, there are invariant subspace Y_1 and Y_2 such that

$$X = Y_1 + Y_2$$

$$\sigma(T|Y_i) \subset G_i,\ i = 1,2.$$

Then $T = T_1 + T_2$ where $T_i = T|Y_i$, $i = 1,2$. Since $0 \in \rho(T_1)$ and $\sigma(T_2) \subset \{0\}$, T_1 is invertible and T_2 is quasinilpotent. Furthermore, by Corollary 1.6, we have

$$Y_1 \cap Y_2 \subset X_T[\sigma(T_1)] \cap X_T[\sigma(T_2)] = X_T[\sigma(T_1) \cap \sigma(T_2)]$$

$$\subset X_T(G_1 \cap G_2) = X_T(\emptyset) = \{0\}.$$

Consequently, $Y_1 + Y_2$ is a direct sum and $T = T_1 \oplus T_2$. □

There is an interesting special case of Theorem 2.1. Let T be a partial isometry with the SDP. If 0 is an isolated point of its spectrum, then T is the direct sum of a unitary and a quasinilpotent operator. This is a special case of the canonical decomposition of a spectral operator in Dunford's sense [7,8]. The spectrum of such a partial isometry has the property

$$\sigma(T) \subset \Gamma \cup \{0\}$$

where Γ is the unit circle on the complex plane.

Such a spectral decomposition of a partial isometry T was obtained in [14] under the asymptotic condition

$$\lim_{n\to\infty} ||T^*T^n - T^nT^*|| = 0$$

where T^* is the Hilbert space adjoint of T. In a finite dimensional space, a partial isometry which is the direct sum of a unitary and a nilpotent operator was obtained in [9] as a consequence of the condition $T^*T^n = T^nT^*$ for some positive integer n.

In a more general case of an operator with a disconnected spectrum, we have:

<u>Theorem 2.2</u> Let $T \in B(X)$ have the SDP and assume that $0 \in \sigma(T)$ and there is an open neighborhood G of 0 such that $\partial G \subset \rho(T)$. Then T is a direct sum of an invertible and a non-invertible operator.

<u>Proof</u> $\{\overline{G}^c, G\}$ being an open cover of $\sigma(T)$, there are invariant subspaces Y_1, Y_2 with the spectral decomposition

$$X = Y_1 + Y_2, \tag{2.1}$$

$$\sigma(T|Y_1) \subset \overline{G}^c, \qquad \sigma(T|Y_2) \subset G.$$

Hence $T_1 = T Y_1$ is invertible and $T_2 = T|Y_2$ is not invertible. Furthermore, with the help of Corollary 1.6 and (1.2), we obtain successively:

$$Y_1 \cap Y_2 \subset X_T[\sigma(T_1)] \cap X_T[\sigma(T_2)] = X_T[\sigma(T_1) \cap \sigma(T_2)] \subset X_T(\overline{G}^c \cap G) = \{0\}.$$

Consequently,

$$X = Y_1 \oplus Y_2 \tag{2.2}$$

and

$$T_1 = T_1 \oplus T_2. \quad \Box \tag{2.3}$$

<u>Theorem 2.3</u> Given $T \in B(X)$ with the SDP, let $G \subset ¢$ be an open neighborhood of $0 \in \sigma(T)$ such that $\partial G \subset \rho(T)$. Then both T and T' admit direct sum decompositions such that each sum has an invertible summand. Specifically,

$$X = X_T(G^c) \oplus X_T(\overline{G}), \tag{2.4}$$

$$X' = X_T(G)^{\perp} \oplus X_T(\overline{G}^c)^{\perp}, \tag{2.5}$$

$T|X_T(G^c)$ and $T'|X_T(G)^{\perp}$ are invertible, and

$$\sigma[T|X_T(G^c)] \subset \overline{G}^c \ , \qquad \sigma[T|X_T(\overline{G})] \subset G, \tag{2.6}$$

$$\sigma[T'|X_T(G)^{\perp}] \subset \overline{G}^c \ , \qquad \sigma[T'|X_T(G^c)^{\perp}] \subset G. \tag{2.7}$$

Proof For the invariant subspaces Y_1 and Y_2 of the spectral decomposition (2.1), property (1.2) gives

$$Y_1 \subset X_T(\overline{G}^c) \subset X_T(G^c) \quad \text{and} \quad Y_2 \subset X_T(G) \subset X_T(\overline{G}).$$

Then (2.1) implies (2.4). T being decomposable (Theorem 1.15), $X_T(G^c)$ and $X_T(\overline{G})$ are spectral maximal spaces and in particular ν-spaces. Consequently, by (1.4), we have

$$\sigma[T|X_T(G^c)] \subset G^c \cap \sigma(T) \subset \overline{G}^c$$

and

$$\sigma[T|X_T(\overline{G})] \subset \overline{G} \cap \sigma(T) \subset G,$$

which proves (2.6).

By Theorems 1.15 and 1.12, T' is decomposable. Thus, for the open cover $\{\overline{G}^c, G\}$ of $\sigma(T') = \sigma(T)$, there are spectral maximal spaces Z_1', Z_2' for T' such that

$$X' = Z_1' + Z_2' \tag{2.8}$$

$$\sigma(T'|Z_1') \subset \overline{G}^c \ , \quad \sigma(T'|Z_2') \subset G.$$

By property (1.2),

$$Z_1' \subset X'_{T'}(\overline{G}^c) \subset X'_{T'}(G^c) \quad \text{and } Z_2' \subset X'_{T'}(G) \subset X'_{T'}(\overline{G}). \tag{2.9}$$

Next, we show that

$$X'_{T'}(G^c) = X_T(G)^{\perp} . \tag{2.10}$$

If $x \in X_T(G)^{\perp}$, then by Corollary 1.14 (1.8), we have

$$\sigma_{T'}(x) \subset \sigma[T'|X_T(G)^{\perp}] \subset G^c$$

and hence $x \in X_{T'}(G^c)$. Thus $X_T(G)^{\perp} \subset X_{T'}(G^c)$. To prove the opposite inequality, we shall follow [16]. Let $x' \in X_{T'}(G^c)$ and $x \in X_T(G)$ be given. Then since $\sigma_T(x) \subset G$, $\rho_T(x)$ and G cover the complex plane.

Define the entire function

$$f(\lambda) = \begin{cases} <\tilde{x}(\lambda), x'> & \text{for } \lambda \in \rho_T(x), \\ <x, \tilde{x}'(\lambda)> & \text{for } \lambda \in G. \end{cases}$$

The function f is well-defined since for $\lambda \in \rho_T(x) \cap G$, we have successively

$$<x,\tilde{x}'(\lambda)> = <(\lambda-T)\tilde{x}(\lambda), \tilde{x}'(\lambda)> = <\tilde{x}(\lambda),(\lambda-T')\tilde{x}'(\lambda)> = <\tilde{x}(\lambda),x'>.$$

Since f is zero at ∞, by Liouville's theorem, we have

$$<x,x'> = \lim_{|\lambda|\to\infty} <\lambda\tilde{x}(\lambda), x'> = 0$$

because $\lambda\tilde{x}(\lambda) = x + T\tilde{x}(\lambda) \to x$ when $|\lambda| \to \infty$.

Thus, $x' \in X_T(G)^{\perp}$ and hence $X'_{T'}(G^c) \subset X_T(G)^{\perp}$. This proves (2.10). Also, by (2.10), we have

$$X'_{T'}(\overline{G}) = X_T(\overline{G^c})^{\perp}. \tag{2.11}$$

Consequently, the spectral decomposition (2.8), with the help of (2.9), (2.10) and (2.11) becomes (2.5). Finally, the inclusions (2.7) follow from Corollary 1.14.

The sums (2.4) and (2.5) are direct for the following reasons:

$$X_T(G^c) \cap X_T(\overline{G}) = X_T(G^c \cap \overline{G}) = X_T(\partial G),$$

$$X_T(G)^{\perp} \cap X_T(\overline{G^c})^{\perp} = X'_{T'}(G^c) \cap X'_{T'}(\overline{G}) = X'_{T'}(G^c \cap \overline{G}) = X'_{T'}(\partial G),$$

and

$$X'_{T'}(\partial G) = X_T(\partial G) = X_T[\partial G \cap \sigma(T)] = X_T(\emptyset) = \{0\}. \quad \square$$

REFERENCES

1 E. Albrecht, "On decomposable operators", Integr. Equations Oper. Theory 2 (1979), 1-10.

2 E. Arghiriade, "Sur l'inverse generalisée d'un opérateur linéaire dans les espaces de Hilbert", Atti Accad. Naz. Lincei Rend.Cl.Fis.Mat.Natur. 45 (1968), 471-477.

3 H. Bart, M.A. Kaashoek and C. D. Lay, "Relative inverses of meromorphic operator functions", Technical Report, University of Maryland, 1974.

4 R. G. Bartle, "Spectral decomposition of operators in Banach spaces", Proc. London Math. Soc. 20 (1970), 438-450.

5 R. G. Bartle and C. A. Kariotis, "Some localizations of the spectral mapping theorem", Duke Math., J. 40 (1973), 657-660.

6. A. Ben-Israel and T. N. E. Greville, "Generalized Inverses: Theory and Applications", John Wiley, New York.

7 N. Dunford, "Spectral theory II. Resolution of the identity", Pacific J. Math. 2 (1952), 559-614.

8 N. Dunford and J. T. Schwartz, "Linear Operators", Part I, Part II, Part III, John Wiley, New York.

9 I. Erdelyi, "Partial isometries defined by a spectral property on unitary spaces", Atti Accad. Naz. Lincei Rend. Cl. Fis. Math. Natur. 44 (1968), 741-747.

10 I. Erdelyi, "A generalized inverse for arbitrary operators between Hilbert spaces", Proc. Camb. Phil. Soc. 71 (1972), 43-50.

11 I. Erdelyi and A. Ben-Israel, "Extremal solutions of linear equations and generalized inversion between Hilbert spaces", J. Math. Anal. Appl. 39 (1972), 298-313.

12 I. Erdelyi and R. Lange, "Operators with spectral decomposition properties", J. Math. Anal. Appl. 66 (1978), 1-19.

13 I. Erdelyi and R. Lange, "Spectral Decompositions on Banach Spaces", Lecture Notes in Mathematics, Vol. 623, Springer-Verlag 1977.

14 I. Erdelyi and F. R. Miller, "Decomposition theorems for partial isometries", J. Math. Anal. Appl. 30 (1970), 665-679.

15 C. Foias, "Spectral maximal spaces and decomposable operators in Banach spaces", Arch. Math. (Basel) 14 (1963), 341-349.

16 S. Frunza, "A duality theorem for decomposable operators", Rev. Roumaine Math. Pures Appl. 13 (1968), 147-150.

17 B. Nagy, "Operators with spectral decomposition property are decomposable", manuscript.

18 J. E. Scroggs, "Invariant subspaces of a normal operator", Duke Math. J. 26 (1959), 95-111.

19 Y. Y. Tseng, "Generalized inverses of unbounded operators between two unitary spaces" (in Russian), Dokl. Akad. Nauk SSSR (N.S.) 67 (1949), 431-434 (reviewed in Math. Rev. 11 (1950), 115).

I. Erdelyi
Department of Mathematics
Temple University
Philadelphia, PA. 19122